MINERALISED PLANT AND INVERTEBRATE REMAINS
A guide to the identification of calcium phosphate replaced remains

Wendy J. Carruthers[1] and David N. Smith[2]
Photography and publication design by James Turner[3]

[1] Sawmills House, Castellau, Llantrisant, CF72 8LQ
[2] Department of Classics, Ancient History and Archaeology, University of Birmingham, B15 2TT
[3] Amgueddfa Cymru - National Museum of Wales, Cathays Park, Cardiff CF10 3NP

Historic England

in collaboration with

UNIVERSITY OF BIRMINGHAM

national museum wales amgueddfa cymru

Published by Historic England, The Engine House, Fire Fly Avenue, Swindon SN2 2EH
www.HistoricEngland.org.uk

Historic England is a Government service championing England's heritage and giving expert, constructive advice, and the English Heritage Trust is a charity caring for the National Heritage Collection of more than 400 historic properties and their collections.

First published 2020

ISBN 978-1-80034-120-3

For more information about images from the Archive, contact Archives Services Team, Historic England, The Engine House, Fire Fly Avenue, Swindon SN2 2EH; telephone (01793) 414600.

Printed and bound by CPI Group (UK) Ltd, Croydon, CR0 4YY

Contents

Introduction

1.1 Scope of the photographic guide

The aim of this photographic guide is to assist archaeobotanists and archaeoentomologists in the identification of mineralised biological remains that are of archaeological significance. In this publication the term 'mineralised' refers to the replacement of biological tissues with calcium phosphate and does not include items preserved by metal corrosion products. Mineralised remains can be difficult to identify because phosphatic mineralisation primarily preserves soft tissues. For fruits and seeds this often results in the loss of diagnostic features of the thickened, protective outer layers (pericarps and seed coats). For insect remains this primarily results in the preservation of fly (Diptera) pupae and puparia which are less familiar to many archaeoentomologists since they mainly study beetles (Coleoptera) preserved by waterlogging. Insects, fruits and seeds can also be preserved as internal casts which are often difficult to identify due to the lack of features.

Widely used reference books are often of limited use for mineralised remains. Fly puparia and the internal structures of fruits and seeds are rarely described or illustrated in sufficient detail for identification purposes. For botanical remains, the dissection of modern reference material is often required (removal of the seed coat or pericarp) before it can be compared to mineralised material. This photographic guide, therefore, provides images (magnifications of x6 to x80) of a range of identifiable plant and insect taxa primarily using mineralised archaeological material, including high magnification images of specific structures and cell layers (magnifications of up to x160). It highlights identification criteria, provides examples of archaeological sites which yielded mineralised material, information on modern ecology and outlines the interpretative value of each taxon.

Fruit and seed anatomy is complex, very variable and little published literature exists describing anatomical details to species or even genus level. The preservation of mineralised fruits and seeds can vary depending on local conditions, as described by McCobb *et al* (2003), with different layers within pericarps and seed coats surviving as the 'surface' (outermost) layer. The guide provides examples of this variation where reference material was available to the authors, although many more variations can be expected to be found in archaeological deposits. Precise descriptions of which cell layers are shown in the photographs have not be given, as detailed anatomical studies would be required for each taxon in order to provide this information.

Insect anatomy and preservation within a species is usually less variable than that seen with plant remains except for minor variations in size. The preservation of insect remains can be exceptional with surface detail being very clear, meaning that many of the same identification features used for modern specimens are available.

This first edition provides information on some of the most commonly found mineralised taxa from cesspits, drains and middens dating from the prehistoric to post-medieval periods. It is envisaged that additional pages will be added, and some pages may be updated, as further well-preserved reference material becomes available. The authors, therefore, would like to hear from colleagues who are willing to lend mineralised botanical and entomological material for photographic purposes. We are particularly eager to receive information about plant and insect remains whose identifications can be confirmed due to the preservation of key characteristics, for example where fruit pericarps are waterlogged and internal structures are mineralised. This volume only contains insects that have been found mineralised from a range of deposits by David Smith or have been published in the wider literature (the exception is the beetle grain pests which have not at this stage been found but it is only a matter of time). As Smith (2013) has shown there is a wide range of insects that could potentially occur in mineralised deposits. Both Girling (1985) and Kenward (1999) have recorded mineralised human ectoparasites from a variety of deposits in York and London. This includes the human flea (*Pulex irritans* L.), the body louse (*Pediculus humanus* L.) and the pubic louse (*Pthirus pubis* L.). Descriptions and illustrations of these finds are included in the respective publications. As more work is undertaken on insect and plant remains from mineralised deposits additional taxa could be incorporated into the guide.

1.2 Organisation of the guide

Section I: Botanical remains
The orientation of the fruits and seeds in the photographs largely follows the most widely used

photographic guide, *Digital Atlas of the Netherlands* (Cappers *et al* 2006) to facilitate comparisons between the two reference resources. Pages are arranged in taxonomic order following Stace (2010), which uses the new Angiosperm Phylogeny Group system of classification at family and genus level.

Section II: Entomological remains
The pages for the insect remains are arranged in the taxonomic order used by Smith (1989).

Section III: Additional items often present in mineralised deposits
A number of other biological (and possible pedological) items in mineralised deposits have been recovered from archaeological samples. While these items can be described and readily recognised they are not always closely identifiable. As these items can provide information about conditions on the site, living conditions, or simply flag up that mineralisation has taken place, they are presented in this section.

Appendix I: Voucher specimens
Information about the botanical and entomological specimens used for photography is provided in this catalogue, arranged numerically by their [B#] and [E#] numbers. The database provides context and site details, linking the specimen to publication information via the site code used in Appendix II. The voucher specimen number is included in the photograph description.

Appendix II: Archaeological sites
This catalogue provides references for archaeological reports used as examples and providing voucher specimens, arranged alphabetically by their county codes (eg, [HA1], [WT1]).

References
Additional cited references are provided, as well as the main bibliographic sources used to provide taxonomic and habitat information.

The main sources consulted on the herbal uses of plants were Grieve 1992 (revised edition, first published in 1931) and Plants for a Future (www.pfaf.org) referenced as 'Web 1'.

The principal references used for information on plant ecology were Stace (2010) and The Online Atlas of the British and Irish Flora (www.brc.ac.uk/plantatlas/)

cited as 'Web 2'.

The principle reference for the ecology of the flies is Smith (1989).

1.3 Sources of reference material and site examples
Most of the reference material used for the botanical and entomological photographs came from either the authors' own reference collections of material retained with the permission of the relevant archaeological organisations, or material loaned by colleagues and organisations specifically for this project. The authors are grateful to the environmental archaeologists and organisations acknowledged below for their help. Botanical [B#] and entomological [E#] 'voucher specimens' are archived, numbered and listed in Appendix I.

At the start of the project the authors used online interest group lists to request information from colleagues about good mineralised archaeobotanical and archaeoentomological assemblages. This research was carried out in order to determine which taxa should be covered in the guide, as well as to provide a range of sites to use as archaeological examples (see Appendix II). The examples used, therefore, are far from exhaustive but do cover sites of different periods and feature types. Geographically most sites are concentrated in the chalklands of Wessex and Hampshire, as might be expected for calcium phosphate mineralisation, though material can become preserved in cesspits and highly organic, moist deposits in all areas of the British Isles, independent of the underlying geology (Green 1979, 281).

2. Phosphatic mineralisation
Since Green (1979) first drew attention to the value of analysing calcium phosphate-replaced plant remains and Skidmore (1999) identified the first set of mineralised insect remains there has been some progress in understanding the process that leads to this type of preservation. Following the excavation of an extensive Late Bronze Age midden-type deposit at Potterne, Wiltshire, in 1983-85 it was recognised that phosphatic mineralisation was not confined to cesspits, latrines, garderobes and drains but, given the correct conditions, could take place *in situ* on a large scale, preserving evidence of the local flora and fauna as well

as deposited biological waste (Carruthers in Lawson 2000). More recent research by Lucy McCobb and Derek Briggs at the University of Bristol has increased our understanding of the conditions required for calcium phosphate replacement and proposed a sequence of events that will result in the phosphatisation of plant remains (McCobb *et al* 2001; McCobb *et al* 2003). The authors refer readers to these papers for a detailed discussion of the process of mineralisation, the first of which discussed the preservation of fruit remains in a 10th century cesspit at Coppergate, York, and the second of which described the replacement of selected seeds and roots at Potterne. To summarise, phosphatisation of a plant or insect remain is dependent on the following key factors (taken from McCobb *et al* 2003):

- The availability of ions from an external source, for example decaying faecal material, animal dung, dumped organic waste.
- The rate of microbial decay in the deposit, water availability and the free movement of pore water, and the pH of the deposit, with acidic products of decay playing a key role in the case of calcium phosphate mineralisation.
- The permeability of plant and insect tissues.

The typical sequence of events leading to phosphatisation involves the smothering of plant and insect material causing death and decay of the remains. Decay causes conditions to turn anoxic and mildly acidic. High concentrations of phosphate and calcium ions in the pore water cause ions to be taken into plant and insect tissues. Where rates of decay are slow the softer tissues (in seeds the endosperm and embryo) take up ions which form the calcium phosphate sub-fossil. Where microbial decay is more rapid structures such as seed coats, where cell walls are often thickened with lignin or tannin, might become at least partially mineralised. Some variations in states of preservation are presented in this guide, for example *Prunus* spp. (pages 8 and 9) where a mineralised whole fruit, a fruit stone and kernels are presented.

3. Acknowledgements

The authors are grateful to the following archaeological organisations for giving us permission to photograph material from their excavations: AOC Archaeology, Canterbury Archaeological Trust, Hereford County Archaeology Service, Network Archaeology, Oxford Archaeology, Pre-Construct Archaeology, University of Leicester Archaeological Services, Wessex Archaeology.

We would also like to thank colleagues who provided us with information about mineralised assemblages, including both grey literature and published reports: Gill Campbell (Historic England), Rachel Fosberry (Oxford Archaeology), Lisa Lodwick (University of Oxford), Angela Monckton (University of Leicester), Ruth Pelling (Historic England) and Wendy Smith (University of Birmingham).

David Smith would like to acknowledge the help of the late Peter Skidmore who kindly identified the first set of voucher specimens from Vine Street back in 1998. He would also like to thank Eva Panagiotakopulu for confirming a number of identifications, particularly that of the *Sepsis* spp.

Wendy Carruthers is very grateful to Wolfgang Stuppy (Ruhr-Universität Bochum, Germany), for generously providing advice on fruit and seed anatomy, though any errors in anatomical descriptions are her own. She would also like to thank Mark Nesbit (Royal Botanic Gardens, Kew) for his very helpful comments and Matt Canti (Historic England) for his advice on mineralised 'nodules'.

We are both grateful to Ruth Pelling (Historic England) and Gill Campbell (Historic England) for encouragement and advice during the production of this guide.

David and Wendy would like to thank James Turner for his tireless work on the project which has gone above and beyond that expected.

Family: DENNSTAEDTIACEAE (in leptosporangiate 'true ferns' group)
Latin name: *Pteridium aquilinum* **(L.) Kuhn**
Common name: bracken
Anatomical element: pinnule

Image description

1. Example 1; pinnule, upper surface [B56].
2. Example 2; pinnule fragment, upper surface [B157].
3. Example 3; upper and lower surfaces [B143].

Key diagnostic features and separation from similar taxa

- Pinnule margin entire (not toothed) and incurled (for the production of spores).
- Broad midrib and clear venation, robust leathery blades.
- Should be separable from other ferns which usually have toothed margins and much weaker venation. However, no mineralised examples of other ferns known to author.

Examples of archaeological sites

Late Bronze Age midden at Potterne [WT1]; Late Iron Age/Early Roman cesspit at Church Street, Maidstone [KT5]; Mid Saxon cesspit at The Deanery, Southampton [HA4].

Modern ecology

Native, common and widespread fern of woods, heathlands, nutrient-poor grasslands and moors, usually on acidic, dry soils. Can dominate large areas.

Interpretative value

In cesspits likely to have been present amongst grassy vegetation used as toilet wipes, to dampen odours, or flooring and roofing materials. In middens may represent deposited waste or vegetation growing locally.

Family: PAPAVERACEAE
Latin name: *Papaver* sp.
Common name: poppy
Anatomical element: seed

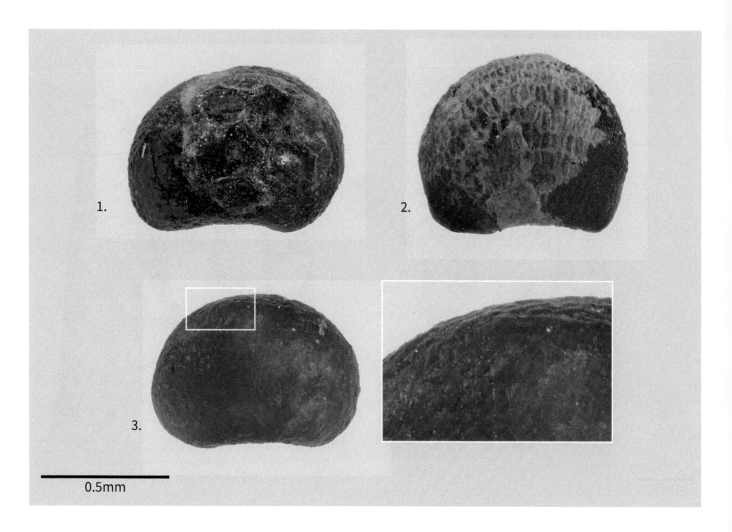

0.5mm

Image description

1. Example 1; seed with some reticulate sculpting of the seed coat preserved; *Papaver* cf. *dubium/ rhoeas* [B28].

2. Example 2; seed, lower layer of seed coat preserved [B99].

3. Example 3; seed, showing cell structure below seed coat [B100].

Key diagnostic features and separation from similar taxa

- Small reniform seeds (generally <1mm) can be identified as *Papaver* sp. from gross morphology and might be identified further if the seed coat is well-preserved.

- Larger seeds may be identifiable as *P. somniferum* if the reticulum is well-preserved (Fritsch 1979).

Examples of archaeological sites

Early Iron Age ditch, Flint Farm, Danebury Environs Project [HA3]; *P. somniferum* from late Roman cesspit at Silchester [HA2]; Anglo-Saxon cesspits, Abbots Worthy [HA7].

Modern ecology

British *Papaver* species are either archaeophytes that grow in arable fields and waste places or introduced species with economic value such as opium poppy.

Interpretative value

Papaver sp. are likely to be food contaminants, seeds used to decorate bread or weeds discarded in cesspits or growing on middens. *P. somniferum* seeds possess only traces of opium alkaloids though they contain a useful oil. Whole capsules (which may retain a few seeds) can be used for medicinal purposes.

Family: FUMARIACEAE
Latin name: *Fumaria* sp.
Common name: fumitory
Anatomical element: seed

0.5mm

Image description

1. Example 1, side and front view [B52].
2. Example 2, side view showing clearer example of depression located opposite apical apertures in pericarp [B123].

Key diagnostic features and separation from similar taxa

- Naked seed usually preserved, without sclerenchymatous pericarp.
- Distinctive features include campylotropous seed (bent longitudinally) and apical depression, as well as gross morphology and cell patterning.
- No similar genera. Unlikely to be identifiable to species level as identification partly relies on features of entire fruits.

Examples of archaeological sites

Late Bronze Age midden at Potterne [WT1]; Iron Age ditch and pits at Battlesbury Bowl [WT2].

Modern ecology

Of ten native and archaeophyte species found in British Isles only two common and fairly widespread; *F. officinalis* and *F. muralis* (Stace 2010; Web 2). All ten species have similar habitat ranges; well-drained arable, waste ground and hedgebanks.

Interpretative value

Used medicinally at least since Roman period, with all parts of plant except roots effective. Main uses of herbal infusions and tinctures were for arthritis, liver and spleen disorders, leprosy, jaundice and skin complaints (Grieve 1992; Web 1). Presence of seeds in the cited examples may be due to plants having grown as weeds, or because fruits present as contaminants of cereal processing waste dumped on the midden-type deposits.

Family: RANUNCULACEAE
Latin name: *Ranunculus acris/bulbosus/repens* - **type**
Common name: buttercup
Anatomical element: seed

1mm

Image description

1. Examples 1 and 2; mineralised seeds. Cell surface observed probably unspecialised layer of seed coat or an inner layer of pericarp [B86] [B32].

2. Example 3, as above with remnant of pericarp [B87]. All three seeds showing slight variations in morphology.

Key diagnostic features and separation from similar taxa

- Flattened seeds with characteristic elongated narrow rows of cells following curved outer edge of seed.

- Size and morphology suggests most mineralised examples come from common terrestrial buttercups such as *R. acris*, *R. bulbosus* and *R. repens*, though some other species may fall into this size range. Without distinctive pericarps these are grouped as 'type'. *R. sardous* can be discounted as seeds have small bumps below spines on the pericarp.

- *Rhinanthus* sp. (page 41) has a similarly flattened seed but is more oval in outline and has distinctive cell pattern.

Examples of archaeological sites

Early Iron Age ditch, Flint Farm [HA3]; medieval deposits in keep, Castle Acre castle [NK1]; 13th/14th century garderobe, Jennings Yard, Windsor [BK1].

Modern ecology

Terrestrial buttercups are often major components of grasslands on a wide range of soils, particularly on grazed pastures. Can also grow as weeds of cultivated land. *R. acris* and R. *bulbosus* often found on dry, nutrient-poor soils while *R. repens* favours more moist, richer ground.

Interpretative value

Likely to be present in cesspits due to the use of hay as toilet wipes, for flooring or to soak up liquids. In middens could have been deposited amongst animal bedding and fodder, or shed from local vegetation. Can also grow as crop weed so could be consumed as contaminant of food.

Family: VITACEAE
Latin name: *Vitis vinifera* L.
Common name: grape
Anatomical element: seed

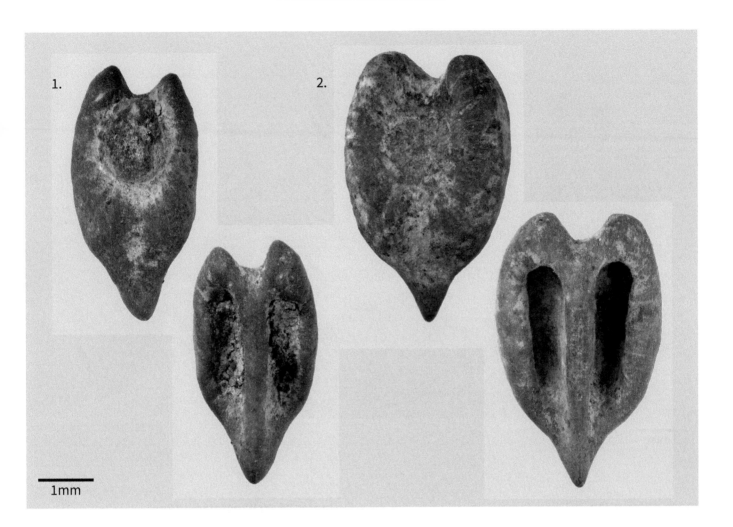

Image description

1. Example 1; dorsal and ventral views [B92].
2. Example 2; dorsal and ventral views [B20].

Key diagnostic features and separation from similar taxa

- Distinctive heart-shaped seed extended into a beak at the base
- Dorsal side has prominent central mound, the 'chalaza scutellum'
- Ventral side with prominent ridge (raphe) with depressions either side (fossettes)
- No similar taxa

Examples of archaeological sites

High medieval to post-medieval cesspit at French Quarter, Southampton [HA8]; post-medieval cesspit and cesstank at St Lawrence Cricket Ground, Canterbury [KT2]

Modern ecology

Cultivated crop, probably originating in Mediterranean basin and south-west Asia (Zohary *et al* 2013).Current growing requirements are for sheltered locations with well-drained, moderately fertile soil, mild winters and enough warmth in summer to ripen fruit (Web 1).

Interpretative value

Often present in urban medieval and post-medieval cesspits. Indicators of faecal material in pits and sewage disposal in drains and rivers. Remains from British faecal deposits most likely from imported grapes though is evidence for cultivation in the medieval period in southern England, particularly at monastic establishments.

Family: FABACEAE
Latin name: *Vicia faba* L.
Common name: field bean
Anatomical element: hila & cotyledons

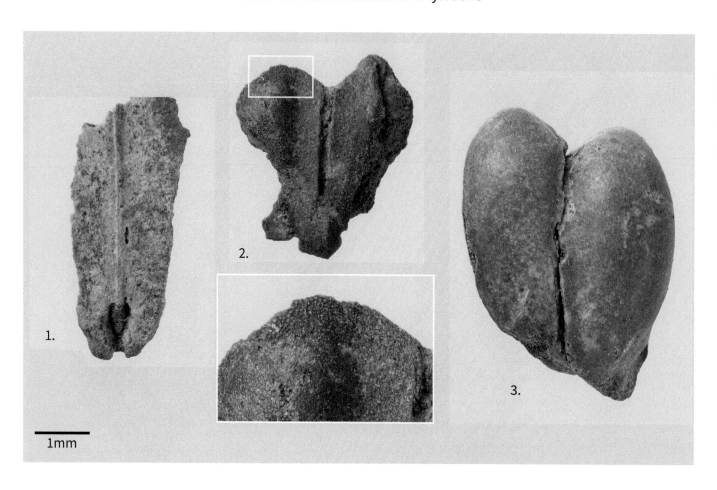

1mm

Image description

1. Example 1; Almost complete hilum with micropyle [B1].
2. Example 2; Fragment of hilum and seed coat [B96].
3. Example 3; Cotyledons and hilar depression, seed coat not preserved [B91].

Key diagnostic features and separation from similar taxa

- Large broadly elliptic seed with much longer hilum than pea (c. 4mm compared with c.1.5mm).
- Some varieties with distinct 'shoulders' at top of seed with hilum in depression. Length and breadth of seed vary considerably depending on variety of bean.
- No similar taxa in this size range.

Examples of archaeological sites

Late Roman cesspit at Silchester [HA2]; middle Saxon cesspits at St Mary's Stadium, Southampton [HA5]; medieval privy at Freeschool Lane, Leicester [LR6].

Modern ecology

Cultivated crop plant present in British Isles from at least the Early Bronze Age (Treasure & Church 2017). Robust annual, moderately frost-tolerant.

Interpretative value

Identifiable fragments recovered from cesspits provide useful information about human consumption of this crop which is under-represented in the charred plant record.

Family: FABACEAE
Latin name: *Pisum sativum* L.
Common name: pea
Anatomical element: seed fragment with hilum; seed

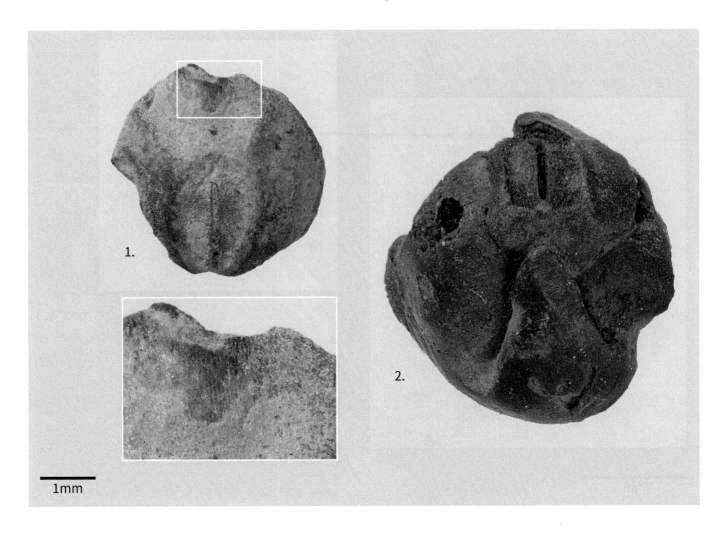

1mm

Image description

1. Example 1; Hilum and surrounding seed coat [B2].
2. Example 2; Seed with some seed coat preserved around hilum [B44].

Key diagnostic features and separation from similar taxa

- Short, oval hilum, large round seed (cf. long hilum of *Vicia faba*, p.6).
- Can only be separated from other pulses where hilum present (or as in Image 2, hilum depression).

Examples of archaeological sites

Middle Saxon cesspits, Southampton [HA4]; Saxon and medieval cesspits, Winchester [HA1]; medieval cesspit, Freeschool Lane, Leicester [LR6].

Modern ecology

Cultivated food plant. Can grow in warm Mediterranean climates as well as cool northern European countries, and is one of the world's most important pulses (Zohary *et al* 2013).

Interpretative value

Usually indicates presence of faecal material, particularly fragments of seed coat (see page 75). Provides valuable evidence of a food rarely identifiable to species-level in charred assemblages.

Family: ROSACEAE
Latin name: *Prunus domestica* L.
Common name: plum (bullace/damson)
Anatomical element: endocarp in concretion and kernel (seed)

1.

2.

5mm

Image description

1. Large *Prunus domestica* cf. *domestica* stone embedded in faecal material (bran curls present) [B41].

2. Incomplete kernel of *Prunus domestica*-type [B164].

Key diagnostic features and separation from similar taxa

- Dimensions and morphology of complete stone comparable with modern reference material.

- Kernel much longer and wider than in *P. spinosa* (c. 11 x 8mm) likely to be *P. domestica*-type but difficult to differentiate between subspecies because of size overlaps and wide morphological variations within group (see *Prunus spinosa* for separation of kernels into 'types').

Examples of archaeological sites

Complete fruit stone from Roman cesspit at 49 St. Peter's Street, Canterbury [KT4]; stones from Late Saxon cesspits at Discovery Centre, Winchester [HA1].

Modern ecology

Cultivated fruit, domesticated from at least the Roman period according to documentary records of planting and grafting (Zohary *et al* 2013). Widely grown from medieval times onwards as orchard fruits.

Interpretative value

Mineralised and part-mineralised/part-waterlogged fruit stones and kernels of *Prunus* sp. can be abundant in cesspits dating from the Roman period onwards. Good indicator of faecal material being present in a feature.

Family: ROSACEAE
Latin name: *Prunus spinosa* L.
Common name: blackthorn or sloe
Anatomical element: whole fruit and kernel (seed)

1.
2.
3.
5mm

Image description

1. *Prunus spinosa* fruit showing mineralised stone (endocarp), flesh (mesocarp) and skin (exocarp) [B40].

2. Mineralised seed removed from waterlogged endocarp which provided confirmation of identification [B168].

3. High magnification image of exocarp [B40].

Key diagnostic features and separation from similar taxa

- Most frequent item preserved is seed which is oval in *P. spinosa* with visible embryo, c.7mm x c.4.5mm (L x W).

- *P. avium* kernel is shorter and more rounded but likely to be some overlap with sloe and other species of cherry.

- Seeds may be tentatively separated into 'types' using length x breadth measurements but likely to be overlaps, particularly in *P. domestica* ssp. *insititia* which is morphologically very diverse.

The following approximate seed dimensions (L x B) were obtained from modern imbibed reference material;

- Sloe/wild cherry-type; oval to round, rounded cross section c. 5.5 to 7.5 x 4 to 5 mm.

- Bullace/damson-type; very variable length/ breadth ratios, flatter cross section, c.8.5 to 12 x 5.5 to 8 mm.

- Plum-type; long and wide, flattened cross section, c.14 to 16.5 x 9 to 10.5mm.

NB. Modern cultivars unlikely to be closely comparable to historic ones as has been much hybridisation and back-crossing with domestic plums (Stace 2010), so seed dimensions provide an approximate guide only.

Examples of archaeological sites

Medieval well re-used for waste disposal at Causeway Lane, Leicester [LR2]; post-medieval cesspit at Bonners Lane, Leicester [LR4].

Modern ecology

Common and widespread native spiny shrub of hedgerows, scrub and woodland, spreading readily by suckers to form dense thickets.

Interpretative value

Used as a food since hunter-gatherer times and commonly found in cesspits, sometimes in large numbers and often in towns. Often preserved with mineralised kernels and waterlogged endocarps in wet faecal deposits. All mineralised *Prunus* species are useful indicators of faecal waste.

Family: ROSACEAE
Latin name: *Malus* sp./*Pyrus* sp.
Common name: apple/pear
Anatomical element: seed

1.

2.

3.

1mm

Image description

1. Example 1; well-preserved, probably partially waterlogged *Malus* sp. seed with elongated cell pattern visible [B46].

2. Example 2; seed with upper and lower seed coat layers visible and base of embryo exposed [B163].

3. Example 3; seed, lowest layer of seed coat preserved showing hilar scar [B167].

Key diagnostic features and separation from similar taxa

- Where surface layers of seed coat preserved elongated, fibre-like cell pattern enables differentiation between *Malus* sp. and small rounded cells of *Pyrus* sp. to be made.

- Where only lower layers identification limited to *Malus* sp./*Pyrus* sp.

- Naked embryo/endosperm very similar in size and appearance to *Prunus spinosa/avium*-type. *Malus* sp./*Pyrus* sp. sometimes thinner, usually slightly flattened on one side.

- Small recognisable fragments of *Malus/Pyrus* sp. seed coat often preserved at the seed apex and around hilum.

Examples of archaeological sites

Late Roman cesspits at Silchester [HA2]; Anglo-Norman pits from French Quarter, Southampton [HA8]; Late 16th to mid 17th century cesspits at Ashmolean Museum, Oxford [OX1].

Modern ecology

Native crab apple (*M. sylvestris*) found in woods, scrub and hedgerows. Difficult to differentiate from cultivated apple and much over-recorded for *M. pumila*, which is often naturalised in hedges, scrub and waste ground (Stace 2010). *Pyrus communis* s.l. is an archaeophyte, probably introduced at early date from central and southern Europe.

Interpretative value

Important indicator of cesspits (Smith 2013). Apples and pears important in human diet in Northern Hemisphere since the Neolithic and cultivated from the Roman period onwards. Used for brewing as well as for a range of medicinal uses, in particular for liver and digestive complaints (Grieve 1992).

Family: ROSACEAE
Latin name: *Rubus* sp.
Common name: bramble
Anatomical element: seed

1mm

Image description

1. Example 1; *Rubus* sp. seed (endocarp not preserved) [B72].

2. Example 2; *Rubus* sect. *Glandulosus* (ID confirmed as waterlogged endocarp removed), face and side views [B73].

3. Example 3; *Rubus* sect. *Glandulosus* (ID confirmed as waterlogged endocarp removed) with high magnification insert showing cell detail [B74].

Key diagnostic features and separation from similar taxa

- Lop-sided oval form, plump in side view, sometimes more angular as in example 1 with undulating surface (possibly not as fully imbibed as other two examples).

- Cell pattern less clear than *Rhinanthus* sp. and *Ranunculus* sp.; convex polygonal cells often giving shiny appearance. More convex in side view and more symmetrical in shape.

- Unlikely to be identifiable beyond *Rubus* sp. when sclerenchymatous endocarp not preserved.

Examples of archaeological sites

Late Bronze Age midden at Potterne [WT1]; late Roman cesspits at Silchester [HA2]; Saxon pits at The Deanery, Southampton [HA4].

Modern ecology

Very common scrambling shrub of waste-ground, scrub, woods and hedgerows. Complex taxonomy due to hybridisation and apomixis (Stace 2010).

Interpretative value

Common in cesspits and middens, often retaining waterlogged endocarp. A useful indicator of faecal deposits where frequent.

Family: ROSACEAE
Latin name: *Agrimonia* **sp.**
Common name: agrimony
Anatomical element: fruit (hypanthium)

1.

2.

1mm

Image description

1. Hypanthium with some preserved bristle bases on distal end. High power magnification showing cell details of woody ridge on hypanthium, with bases of fine and coarse hairs producing verrucose surface [B47].

2. Specimen from distal end showing bases of bristles [B47].

Key diagnostic features and separation from similar taxa

- Distinctive fruit not readily confused with other taxa, especially where bristles preserved.

- Difficult to separate two British species, *A. eupatoria* and *A. procera* unless complete. Small differences in size, morphology, bristles and extent of ridging given in Stace (2010).

Examples of archaeological sites

Late Saxon cesspit from Discovery Centre, Winchester [HA1].

Modern ecology

Perennial herbaceous plant of basic/neutral soils in hedgebanks, waysides, woodland margins, field borders, open grassland and sometimes waste ground. Most often found on dry, nutrient-poor soils.

Interpretative value

Medicinal uses include herbal teas for the control of diarrhoea, skin, mouth and pharynx complaints (Grieve 1992; Web 1). In faecal deposits possibly an accidental inclusion as contaminant of cereal-based foods. Hooked fruits may be carried into latrines on clothes or amongst materials used for flooring, roofing or toilet wipes.

Family: ROSACEAE
Latin name: *Aphanes* **sp.**
Common name: parsley-piert
Anatomical element: seed

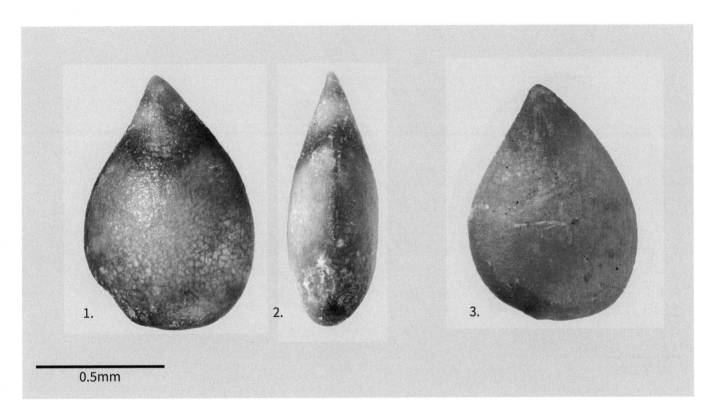

0.5mm

Image description

1. Example 1; face view; seed, pericarp not preserved, showing irregular polygonal cells of storage tissues [B110].
2. Example 1; side view [B110].
3. Example 2; face view; seed with traces of pericarp preserved (bottom left), horizontally elongated cells giving finely banded appearance [B6].

Key diagnostic features and separation from similar taxa

- Small seed with distinctive laterally compressed drop-shaped form, acentric position of basal scar and apex slightly curved towards same side.
- Two British species, *A. arvensis* and *A. australis* produce very similar seeds so unlikely to identify to species level.
- Broader and more flattened than *Alchemilla* sp. with more curve in apex. More acute apex, compressed seed and difference in position of attachment scar compared with *Potentilla* sp. and *Fragaria vesca*.

Examples of archaeological sites

Iron Age ditch and pits at Battlesbury Bowl [WT2].

Modern ecology

Both species are winter (less often spring) germinating annuals of arable, open grasslands and bare patches on dry basic or acidic soils. Very similar habitat ranges but *A. australis* usually replaces *A. arvensis* on more acidic, less fertile soils (Web 2).

Interpretative value

At Battlesbury most likely growing on dry, chalky ditch and pit margins, shedding seeds into midden-type deposits. May also have grown as arable weed and been deposited amongst processing waste. Very little evidence of faecal waste at this site but could be present as food contaminant.

Family: MORACEAE
Latin name: *Ficus carica* L.
Common name: fig
Anatomical element: fruit & seed

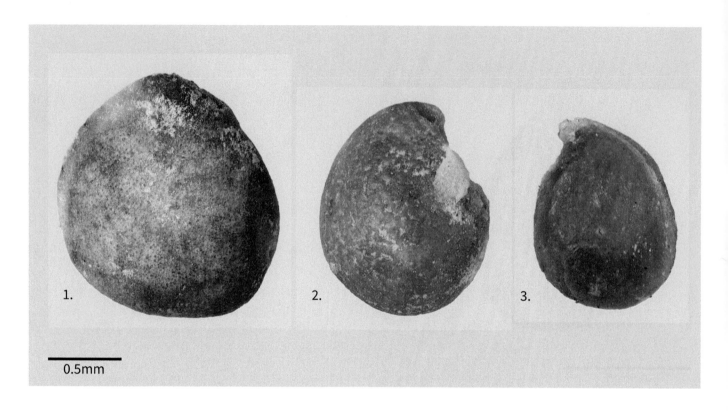

0.5mm

Image description

1. Example 1; fruit, pericarp preserved [B132].
2. Example 2; seed preserved only, no pericarp [B63].
3. Example 3; smaller fruit demonstrating wide size range and some variation in morphology within species [B35].

Key diagnostic features and separation from similar taxa

- Small, broadly ovate fruits with ridge along one side leading to attachment scar just below apex.
- Some variation in size and shape depending on position in syconium.
- Broader at base, without curved fin-like apex and lacking basal attachment scar of *Aphanes* sp.
- Lacking rounded 'beak' and surface ridges of *Potentilla* sp. and *Fragaria* sp. fruits.

Examples of archaeological sites

Late medieval stone-lined cesspits, St Nicholas Place, Leicester [LR5]; late 16th-mid 17th century cesspit/cellar backfill, Ashmolean Museum, Oxford [OX1].

Modern ecology

Cultivated food originating from Mediterranean basin and south-west Asia (Zohary *et al* 2013). On British sites most likely to have been imported, although cultivation is possible.

Interpretative value

Commonly recovered from cesspits, river sediments etc., mainly in urban locations dating to Roman and medieval periods. One of best indicators for identification of human faecal material as non-native species with seeds too small to be spat out (Smith 2013). As well as indicating probable importation of food, has medicinal uses, including as laxative and to ease catarrhal infections (Grieve 1992).

Family: URTICACEAE
Latin name: *Urtica dioica* L.
Common name: common nettle
Anatomical element: seed

0.5mm

Image description

1. Examples 1 & 2, seed, pericarp not preserved [B8] [B103].
2. Example 3; seed, traces of pericarp/seed coat preserved [B102].

Key diagnostic features and separation from similar taxa

-   Small elliptic seeds very similar to modern reference specimens, though pericarp usually not preserved resulting in slightly cordate base.
-   *U. urens* much larger and more ovate in outline (see page 16).

Examples of archaeological sites

Late Bronze Age midden, Potterne [WT1]; Iron Age ditches and pits at Battlesbury Bowl [WT2]; late Roman cesspits at Silchester [HA2].

Modern ecology

Widely distributed perennials of woods, fens, waysides, cultivated land and waste places. Prefers damp, nutrient enriched soils (primarily high phosphorus), eg around human occupation.

Interpretative value

Can be abundant in middens, probably deriving from local vegetation. In faecal deposits may originate from plants growing locally on enriched soils or ingested as crop contaminant. Consumed as a leaf vegetable in the past and whole plant used medicinally for wide range of complaints, including rheumatism, sciatica and improving circulation (Phillips 1983, 18).

Family: URTICACEAE
Latin name: *Urtica urens* L.
Common name: small nettle
Anatomical element: seed

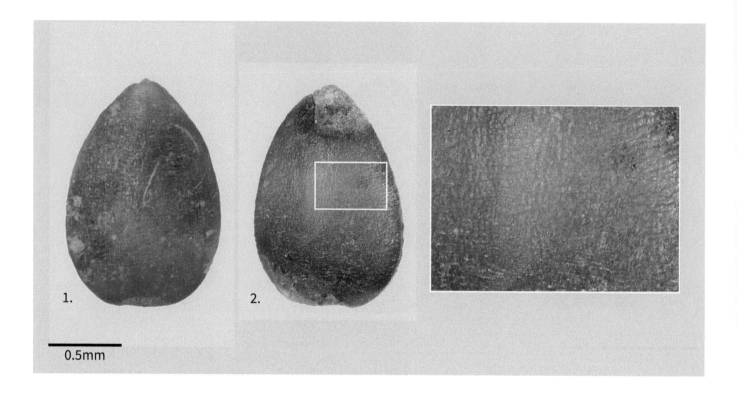

0.5mm

Image description

1. Example 1, seed, pericarp not preserved [B10].
2. Example 2, seed, pericarp not preserved [B104].

Key diagnostic features and separation from similar taxa

- Ovate seed with polygonal cell pattern of endosperm (see McCobb *et al* 2003, Fig.1). Gross morphology compares closely with modern reference material.
- Occasionally areas of thin seed coat preserved
- *U. urens* is much larger than *U. dioica* and widest point of seed is closer to the base. Similar cell pattern of endosperm usually visible in both species.

Examples of archaeological sites

Late Bronze Age midden at Potterne [WT1]; Iron Age midden deposits at Battlesbury Bowl [WT2]; Early Iron Age ditch at Flint Farm, Danebury Environs Project [HA3]; earlier medieval well and latrine pit at Causeway Lane, Leicester [LR2].

Modern ecology

Annual archaeophyte of cultivated and waste ground. Prefers light, sandy soils with high phosphate levels (Pigott & Taylor 1964). Nitrophilous plant and indicator of lime deficient soils, tolerant of heavy shading (Greig-Smith 1948).

Interpretative value

Because of preference for nutrient-rich soils likely to be found growing close to middens and features containing faecal waste. Can also be consumed as a leaf vegetable and whole plant used for medicinal purposes, including for gout, urticaria and as anti-inflammatory. Grieve (1992, 578) notes old herbals recommend ingestion of seeds for stings, venomous bites, and as antidote to poisoning by hemlock, henbane and nightshade. Seeds also used medicinally for consumption, fevers and goitre.

Family: VIOLACEAE
Latin name: *Viola* **L.**
Common name: violet or pansy
Anatomical element: seed

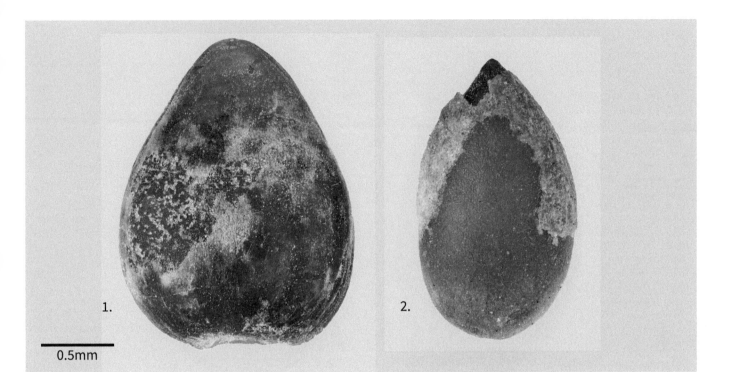

1.

2.

0.5mm

Image description

1. Example 1; seed (slightly compressed) with areas of smooth rectangular cells of seed coat preserved [B138].

2. Example 2; small area of seed coat poorly preserved, polygonal cell pattern of endosperm and embryo exposed [B148].

Key diagnostic features and separation from similar taxa

- Main features are a prominent basal scar where seed coat preserved, sometimes present as depression (as in Example 1), circular cross-section, smooth shiny seed coat with longitudinal rectangular cells, slight depression on one side near apex where hilum and caruncle (oily outgrowth to lure ants) would be located in intact seed. The last characteristic not always clear in mineralised seeds without seed coat.

- In British Isles thirteen species shows some variation in length/breadth ratios and size but share main morphological characteristics.

- No other taxa share all features.

Examples of archaeological sites

Iron Age ditch and pits at Battlesbury Bowl [WT2]; medieval deposits in Castle Keep, Castle Acre Castle [NK1].

Modern ecology

Annuals and perennials. Native and archaeophyte species grow in wide range of habitats, including cultivated and waste ground, grassland, hedgebanks, woods. More local species are found on scrub, heaths, moors, bogs, upland rocky places and by the sea.

Interpretative value

Limited interpretative value due to wide habitat range. Most likely present amongst grassy vegetation used for floor covering, toilet wipes and to absorb moisture. However, Grieve (1992) lists wide range of medicinal uses listed for sweet violet (*V. odorata*) in herbals where all of plant or dried flowers used in infusions and distillations. Medicinal properties include use as laxative, treatments for fevers, epilepsy, inflammation, pleurisy, jaundice and quinsy. Romans used flowers to make wine. Fresh sweet violet flowers provide perfume, and sugared violet flowers were used in conserves.

Family: LINACEAE
Latin name: *Linum catharticum* L.
Common name: purging flax
Anatomical element: seed

1.

2.

0.5mm

Image description

1. Example 1; slightly convex side, seed coat not preserved, polygonal cells of endosperm/embryo visible [B30].
2. Example 2; flatter side of seed, no seed coat, polygonal cells of endosperm/embryo [B16].

Key diagnostic features and separation from similar taxa

- Characteristic *Linum* sp. asymmetrical, laterally compressed elliptic shape but much smaller in size than other species (c. 1.25 mm).
- More regular rounded cells of seed coat, aligned around edges of seed not preserved.
- No similar seeds of this size range and morphology.

Examples of archaeological sites

Early and middle Iron Age pits and ditch at Battlesbury [WT2]; early Iron Age ditch at Flint Farm, Danebury Environs Project [HA3]; Anglo-Saxon sunken-featured building at Abbots Worthy [HA7].

Modern ecology

Annual or biennial, on infertile dry calcareous or base-rich grasslands but also in mires and flushes on neutral to mildly acidic soils, in short-sedge fen and dry heaths. Widespread in British Isles.

Interpretative value

May be incidental inclusion in grassy material discarded in midden-type waste but also useful medicinally as purgative, for liver complaints etc. so possibly consumed.

Family: LINACEAE
Latin name: *Linum usitatissimum* L.
Common name: flax
Anatomical element: seed

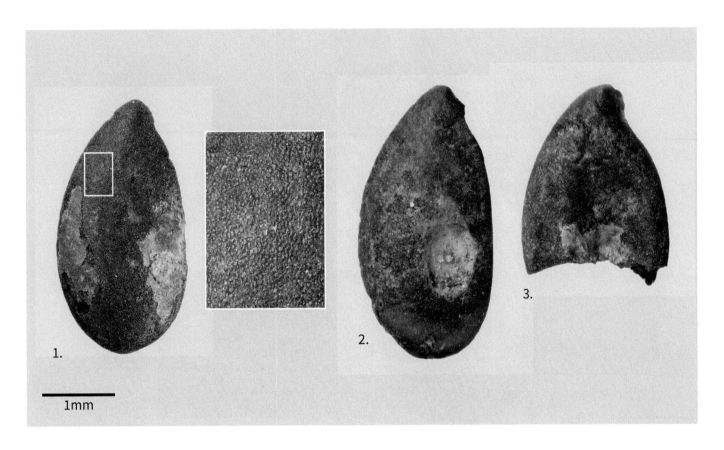

Image description

1. Example 1; showing preservation of traces of seed coat and high magnification detail of cells of endosperm/embryo [B93].

2. Example 2; seed coat not preserved [B94].

3. Example 3; half seed with well-preserved apex, traces of seed coat only, *Linum* cf. *usitatissimum* [B95].

Key diagnostic features and separation from similar taxa

*   When complete or near complete, large size and gross morphology provide sufficient characteristics to separate from smaller three British species of *Linum*, although there may be overlap with fragments of *L. perenne*.

*   Cell size varies across seed coat, with larger cells in centre. Where sufficient seed coat is preserved this character separates *L. usitatissimum* from other British species.

*   No similar large flattened asymmetric seeds outside this genus.

Examples of archaeological sites

Late Bronze Age midden-type deposit at Potterne [WT1]; medieval (AD 1250-1400) cesspit near Ferrer's Road, Huntingdon [CB1].

Modern ecology

Annual, self-pollinated cultivated crop plant grown for fibre and oil since the Neolithic in Europe and western Asia (Zohary *et al* 2013). Today grows as a casual from crops and waste birdseed on field margins and waste ground. Flax grows best on well-drained deep soils of average fertility in sunny location. Plants are grown for production of oil, fibre and seeds.

Interpretative value

Seeds likely to be present in cesspits and middens through being consumed either for medicinal reasons or as flavourings and decoration. Medicinal uses include external use in poultices for the reduction of inflammation and irritation, and internal use as a laxative (Grieve 1992).

Family: BRASSICACEAE
Latin name: *Brassica/Sinapis* **sp.**
Common name: cabbages/mustards etc
Anatomical element: seed

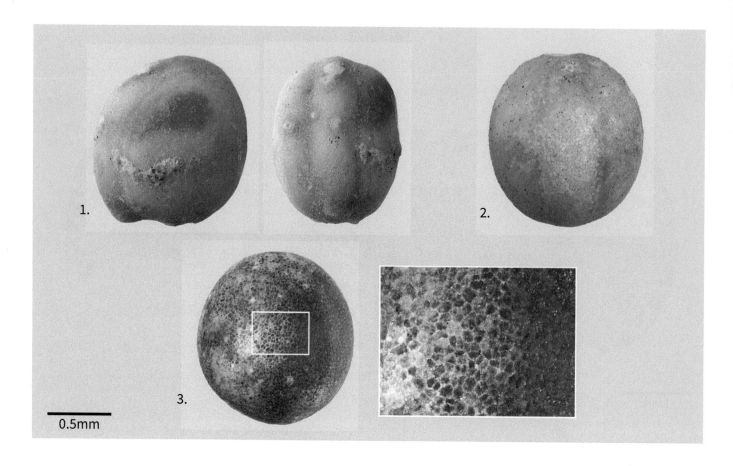

0.5mm

Image description

1. Example 1, side and front view; embryo curled around storage tissues clearly visible [B120].
2. Example 2; more rounded seed, embryo faintly visible [B36].
3. Example 3; slightly larger seed, cell pattern clearer against darker seed [B121].

Key diagnostic features and separation from similar taxa

- Most seeds evenly rounded with clear irregular polygonal cell pattern of endosperm/embryo.
- No other perfectly rounded seeds in c. 1 – 1.5 mm size range. Small *Vicia* sp. unlikely to become mineralised and would have visible radicle and no curved embryo.

Examples of archaeological sites

Very large number in Iron Age ditch and pits at Battlesbury Bowl [WT2]; mid-Saxon pits at Anderson's Road, Southampton [HA9].

Modern ecology

Taxon encompasses wide range of weeds and economically useful plants, including black mustard, cabbage, turnip, white mustard, charlock. Some possibly native in coastal regions, most are archaeophytes, introduced, often naturalised. Charlock (*Sinapis arvensis*) is common weed of arable and waste ground.

Interpretative value

Brassica/Sinapis sp. is the most frequently mineralised type of plant remain, particularly in calcareous areas even where no apparent faecal or midden-type deposits present, and sometimes present in large numbers, as at Battlesbury Bowl. Because not closely identifiable, of little value unless found in large concentrations in faecal deposits or middens, where could represent use as mustard, as oil seed crop or for medicinal purposes. Used externally in poultices, internal use of mustard powder as an emetic, for toothache and as an antiseptic (Grieve 1992).

Family: BRASSICACEAE
Latin name: *Barbarea/Sisymbrium*–type
Common name: winter-cress/rocket-type
Anatomical element: seed

0.5mm

Image description

1. Example 1; seed, cell structure poorly preserved [B49].

2. Example 2; seed with seed coat not preserved, cells of endosperm/embryo well preserved [B156].

Key diagnostic features and separation from similar taxa

- Distinctive angular oblong outline with indentation at hilar end. Cotyledon side of seed (left) much wider than radicular (right) (Berggren 1981, 139 & 141).

- *Barbarea vulgaris* and larger seeds of *Sisymbrium officinale* with seed coats removed bear closest resemblance to archaeological specimens examined.

- Lack of diagnostic characters of seed coat means some other genera from the Brassicaceae could fall into this group if similar length/breadth ratio and asymmetric form, eg *Nasturtium* sp. (NB. Removal of seed coat reduces length, and seeds vary considerably in size and form in most genera with this seed morphology).

Examples of archaeological sites

Late Bronze Age midden-type deposit at Potterne, Wilts [WT1].

Modern ecology

The only native and most widespread species of *Barbarea* sp., *B. vulgaris*, grows in damp places by rivers and ditches, but also drier disturbed places and in nutrient-enriched soils (Ellenburg 1988). *Sisymbrium officinale* is annual/biennial archaeophyte of cultivated and waste places, mainly on dry, open, base-rich or neutral soils.

Interpretative value

B. vulgaris likely to have grown on highly organic, moisture-retentive deposits such as middens or around cesspits. Freshly harvested *S. officinale* plants are rich in vitamin C and were valued in ancient times for treating colds and throat complaints (Grieve 1992; Web 1). Seeds used to make mustard (Web 1).

Family: BRASSICACEAE
Latin name: *Thlaspi arvense* **L.**
Common name: field penny-cress
Anatomical element: seed

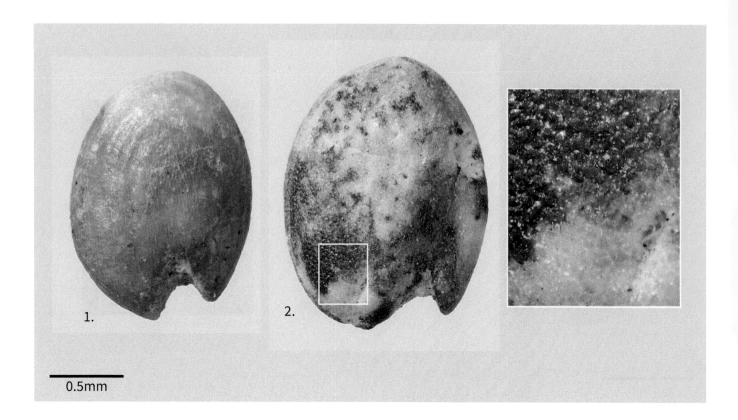

0.5mm

Image description

1. Example 1; seed with at least upper layers of seed coat not preserved but slight impression of ribbed seed coat retained [B9].

2. Example 2; seed with seed coat not preserved, polygonal cells of endosperm/embryo visible [B122].

Key diagnostic features and separation from similar taxa

- Elliptic outline with deep notch at hilum end due to absence of seed coat. Identification more secure where impressions of narrow ribs on seed coat following seed outline are visible.

- Without impressions of ribs may not be separable from a number of other similar sized and shaped members of Brassicaceae eg *Lepidium* sp. Identification as '*Thlaspi*-type' may then be appropriate.

Examples of archaeological sites

Iron Age ditch and pits at Battlesbury Bowl [WT2]; medieval cesspits at Causeway Lane, Leicester [LR2]

Modern ecology

Annual archaeophyte of arable and broad-leaved crops, waysides, waste ground and a garden weed, mainly found on heavier, moist soils in open locations.

Interpretative value

Likely to be present as a contaminant of cereal-based foods or amongst vegetation collected as toilet wipes, flooring etc. May have been growing near to, or smothered by, midden deposits at sites such as Battlesbury Bowl or present amongst discarded crop processing waste. Grieve (1992) lists seeds as an ingredient of 'Mithridate', ancient concoction said to have been an antidote to poisoning, a version of which was used up to the 19th century.

Family: POLYGONACEAE
Latin name: *Fallopia convolvulus* (L.) Á. Löve
Common name: black bindweed
Anatomical element: fruit

1mm

Image description

1. Example 1; whole fruit, possibly less imbibed than example 2 or not as mature as narrower and sides more concave [B12].

2. Example 2; whole fruit and high magnification image showing cell details of pericarp [B124].

Key diagnostic features and separation from similar taxa

- Large elliptic-rhombic trigonous nutlet with rough surfaces between three edges.

- Closely resembles modern reference material due to preservation of pericarp. Naked seeds could be found in archaeological deposits but have not yet been observed by author.

- Distinctive in size and gross morphology. *F. dumetorum* has smaller, smooth fruits and *Fagopyrum esculentum* has much larger, slightly winged fruits.

Examples of archaeological sites

Late Bronze Age midden at Potterne [WT1]; medieval cesspit at Stour Street, Canterbury [KT1].

Modern ecology

Widespread and common annual archaeophyte of arable, gardens, roadsides and waste ground.

Interpretative value

Most likely present amongst crop processing waste or as a weed growing close to middens and cesspits. May have been present in cesspits amongst grassy vegetation used as toilet wipes, floor covering, to soak up liquids or reduce odours.

Family: POLYGONACEAE
Latin name: *Rumex acetosella* L.
Common name: sheep's sorrel
Anatomical element: seed

0.5mm

Image description

1. Seed, edge to front, with pericarp and embryo not preserved. Embryo groove to left [B17].
2. Seed, face to front showing embryo groove, embryo and pericarp not preserved [B17].

Key diagnostic features and separation from similar taxa

- Very small, rounded trigonous seed with embryo not preserved in this example.
- Due to small size and very rounded but trigonous form, not easily confused with other taxa.

Examples of archaeological sites

Present in low numbers in all levels of Late Bronze Age midden at Potterne [WT1]; single seeds present in two Iron Age pits at Battlesbury Bowl [WT2]

Modern ecology

Native rhizomatous perennial of short grasslands on impoverished, acidic sandy or stony soils. Also found on heaths and coastal locations such as sand dunes.

Interpretative value

At Battlesbury Bowl, located on Cretaceous chalk, these remains most likely represent vegetation brought in from heathland, or possibly dung from livestock that had grazed on poor grasslands away from the site. In cesspits could be present in grassy vegetation used as toilet wipes or for flooring.

Family: POLYGONACEAE
Latin name: *Rumex cripsus*-type.
Common name: curled dock-type
Anatomical element: seed

1mm

Image description

1-3. Examples showing variations in length/breadth ratios, pericarps not preserved. Paler, slightly raised embryo visible in middle of left face of example 3 [B18] [B90] [B88].

4. Example 4; pericarp and embryo not preserved [B89].

Key diagnostic features and separation from similar taxa

- Elliptic-rhombic outline with characteristic trigonous cross section.

- Embryo sometimes visible, positioned along the middle of one of the three sides.

- This 'type' includes *Rumex* species within the 1.5 to 2.5mm length size range, taking into account absence of pericarp; *R. crispus; R. conglomeratus; R. sanguineus; R. obtusifolius; R. acetosa; R. palustris. Polygonum aviculare* is distinctive in being asymmetrically trigonous. Trigonous examples of *Persicaria* sp. are either larger or more flattened and asymmetrical in cross-section.

- Although some species can be ruled out due to obvious size differences, mineralised seeds in this group are unlikely to be identifiable to species level due to size overlaps, possibility of dimensions changing depending on how imbibed the seed was at time of mineralisation, as well as absence of characters of pericarp.

Examples of archaeological sites

Rumex sp. identified from Early Iron Age ditch at Flint Farm [HA3]; *Rumex* sp. common in Iron Age ditch and pits at Battlesbury Bowl [WT2].

Modern ecology

Group includes majority of common species found in British Isles, most of which are primarily plants of grassy, disturbed or cultivated land. Habitats also include damp, nutrient-rich ground near water (*R. palustris*) and shady, damp places such as woods and hedgerows (*R. sanguineus*). All except *R. acetosa* show preference for nutrient-enriched soils.

Interpretative value

Most likely to represent vegetation growing close to or on midden deposits and by ditches containing richly organic waste. Not common in cesspits but could be present amongst materials used as toilet wipes, flooring etc. or accidentally ingested as crop contaminant. Limited interpretative value due to group having a wide habitat range.

Family: CARYOPHYLLACEAE
Latin name: *Stellaria* sp.
Common name: stitchwort
Anatomical element: seed

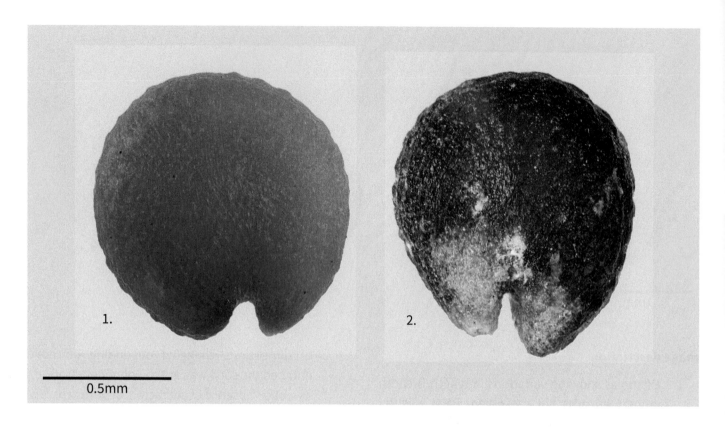

0.5mm

Image description

1. Example 1; seed, seed coat not preserved, though low, sometimes pointed, mounds present. Tip of radicle and cotyledons equal in length (Berggren 1981, 56). Identified as *Stellaria* cf. *media* on basis of size and characters [B22].

2. Example 2; seed, seed coat not preserved but rows of longer curved mounds present. Tips of radicle and cotyledons uneven in length (Berggren 1981, 57). Identified as *Stellaria* cf. *graminea* on basis of size and characters [B105].

Key diagnostic features and separation from similar taxa

- Size and symmetry of rounded, notched seed are key features, also type of verrucose ornamentation, though much less distinct than in seeds with intact seed coats.

- Tentative identification to species level may be possible, or to 'type' depending on preservation. Possibility of some overlap in size and form with other species and genera, eg *S. nemorum, Myosoton aquaticum, Cerastium tomentosum.*

Examples of archaeological sites

Late Bronze Age midden at Potterne [WT1]; Anglo-Saxon sunken-featured building at Abbots Worthy [HA7]; medieval cesspits at Northgate House, Winchester [HA1].

Modern ecology

S. media is rapid coloniser of a range of disturbed habitats capable of two or three generations in a year. Widely found as garden weed, weed of cultivated land and on nutrient-rich habitats such as compost heaps. *S. graminea* is perennial of neglected pastures and woodland clearings. Other taxa mentioned occupy habitats ranging from damp/wet places to waste ground.

Interpretative value

Likely to have colonised bare or disturbed soil around nutrient-rich middens and cesspits. Whole plant is greatly valued by herbalists both for external use in ointments to reduce swelling and treat abscesses and internally in decoctions for coughs, constipation and scurvy amongst other ailments (Grieve 1992).

Family: CARYOPHYLLACEAE
Latin name: *Agrostemma githago* L.
Common name: corncockle
Anatomical element: seed and seed coat impression

1mm

Image description

1. Side view of seed, thickened outer layers of seed coat not preserved [B11].
2. Impression of seed coat in faecal material (cereal bran curls visible within concretion) [B65].

Key diagnostic features and separation from similar taxa

- Large laterally compressed, lop-sided reniform seed closely comparable to reference specimens, although spinose ornamentation reduced to low mounds.

- Impressions in faecal material show distinctive regular rows of voids where the spinose seed coat has become embedded and decayed.

- Separated from other members of the Caryophyllaceae due to large size and distinctive morphology. With impressions the size and spacing of the depressions are important to separate from other papillose seeds eg *Silene* sp.

Examples of archaeological sites

Widely found in urban faecal deposits, particularly Anglo-Saxon and medieval cesspits, eg Saxon and medieval pits in Winchester [HA1]; Saxon pits at The Deanery, Southampton [HA4].

Modern ecology

Archaeophyte, annual weed of arable and disturbed places, formerly common and widespread, now rare as arable weed in British Isles.

Interpretative value

Indicator of faecal material, particularly seed coat impressions which may originate from contaminants in flour. Large numbers of whole seeds may suggest the deposition of waste from crop cleaning. Although poisonous in high concentrations some past medicinal uses have been suggested (Grieve 1992).

Family: CARYOPHYLLACEAE
Latin name: *Silene* sp.
Common name: campion
Anatomical element: seed

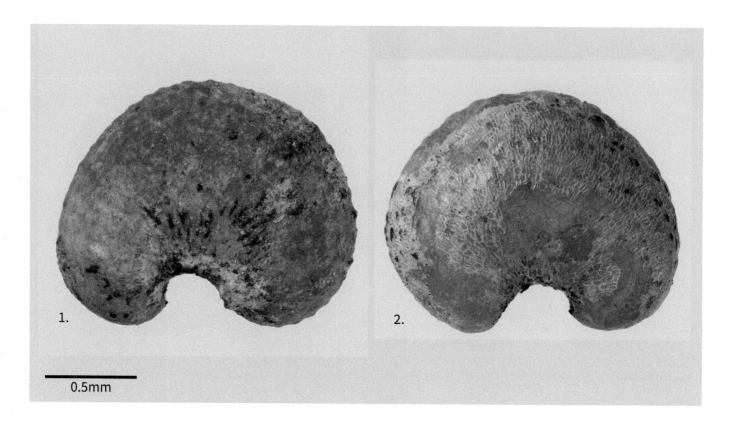

0.5mm

Image description

1. Example 1; seed, traces of seed coat preserved [B13].
2. Example 2; seed, traces of seed coat preserved [B154].

Key diagnostic features and separation from similar taxa

- Reniform or notched semi-circular seed with variable traces of papillose seed coat surviving.
- Where ornamentation of seed coat not adequately preserved, identification beyond genus unlikely due to overlaps in sizes and absence of seed coat characters.

Examples of archaeological sites

Late Bronze Age midden at East Chisenbury [WT3]; Anglo-Saxon cesspit at Abbots Worthy [HA7]; medieval cesspit at Stour Street, Canterbury [KT1].

Modern ecology

Native species and archaeophytes range from plants growing in open grasslands, cultivated soils, waste ground and waysides on well-drained soils (*S. vulgaris*; *S. gallica*; *S. latifolia*; *S. noctiflora*) to plants that can tolerate deep shade in hedgerows and woodlands (*S. dioica*)

Interpretative value

Occasionally recovered from middens and cesspits located on base-rich soils. Most likely present as contaminants of cereal-based foods or plants growing close to middens and waste pits. May also be gathered amongst grassy materials being used as toilet wipes, floor covering etc.

Family: AMARANTHACEAE
Latin name: *Chenopodium/Atriplex* -type
Common name: fat hen/orache-type
Anatomical element: seed

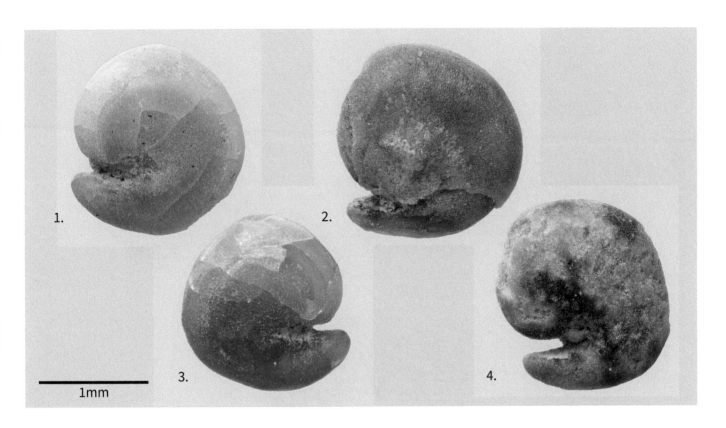

1mm

Image description

1-4. Four examples of seeds without seed coats showing variations in form and size. Partial preservation of seed coat in example 2 only [B81] [B80] [B85] [B82].

Key diagnostic features and separation from similar taxa

- Curved embryo around storage tissues (mainly perisperm) typical of family.

- Wide variation in size and to some extent morphology in *Atriplex* spp.; *Chenopodium* spp. more evenly rounded but likely to be overlaps between two genera. These four seeds more typical of *Atriplex* sp. in irregular outlines and large size.

- Some overlap in size and gross morphology possible with coatless seeds of other members of the family, eg *Suaeda maritima*, hence use of 'type'.

Examples of archaeological sites

Abundant in Late Bronze Age midden at Potterne [WT1]; early Iron Age ditch at Flint Farm, Danebury Environs Project [HA3]; middle/late Iron Age pits at Elms Farm, Leicester [LR1].

Modern ecology

The most common species in group are native and archaeophyte annuals, typically found as weeds of cultivated and waste ground, especially where nutrient levels raised by manuring.

Interpretative value

Can be present in large numbers in middens, probably due to colonisation of nutrient-rich deposits. Present but less abundant in cesspits, possibly having grown nearby or being gathered amongst grassy vegetation used for flooring and toilet wipes. However, several members of Chenopodiaceae also edible, eaten as seeds and leaf vegetable by various cultures today. Medicinal uses for *C. album* and *C. bonus-henricus* include to aid digestion, as laxative and vermifuge, and externally to assist healing (Web 1).

Family: PORTULACACEAE
Latin name: *Montia fontana* **L.**
Common name: blinks
Anatomical element: seed

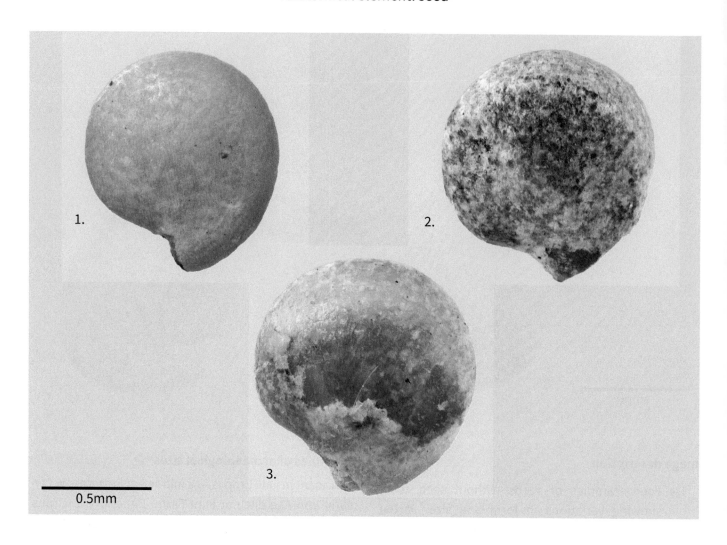

0.5mm

Image description

1. Example 1; seed, lignified seed coat not preserved, tubercles reduced to low mounds [B4].

2-3. Examples with seed coats not preserved, apart from possible area at base of example 3 [B118] [B119].

Key diagnostic features and separation from similar taxa

- Raised tubercles visible on the best preserved examples, but faint in some cases.

- Small size and very compact, rounded form distinguishes *Montia* sp. from more angular tubercled reniform Caryophyllaceae (see *Stellaria*-type, page 26) and similar shaped smooth Chenopodiaceae (see *Chenopodium/ Atriplex*-type, page 29).

Examples of archaeological sites

Late Bronze Age midden-type deposit at Potterne [WT1].

Modern ecology

Native, annual to perennial, damp to wet places eg seasonally damp hollows, flushes, waterside. Widespread in British Isles.

Interpretative value

Useful indicator of vegetation growing on open, damp soils. Not yet recovered from faecal deposits according to published literature. Relatively high frequency at Potterne thought to be due to impeded drainage in the basal layers of the deposit, with the formation of a mineralised 'hard-pan' basal layer in one area of the site.

Family: PRIMULACEAE
Latin name: *Anagallis arvensis*-**type**
Common name: scarlet pimpernel-type
Anatomical element: seed

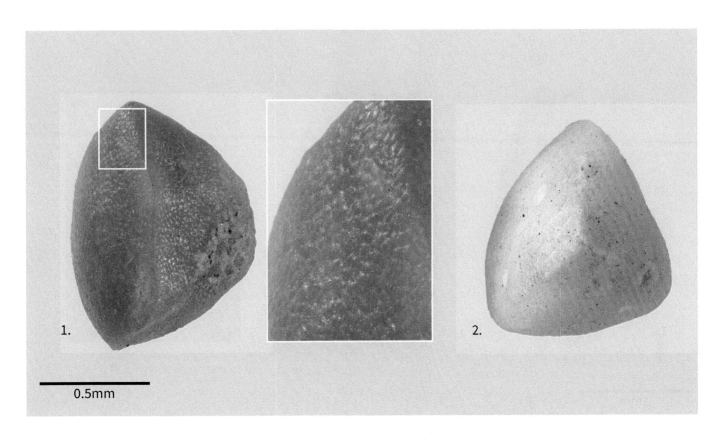

1.

2.

0.5mm

Image description

1. Example 1; seed, side view, seed coat not preserved [B19].

2. Example 2; seed viewed from above, seed coat not preserved [B61].

Key diagnostic features and separation from similar taxa

- Polygonal pyramid shaped seed with clear pattern of small, polygonal cells. Some variations in form, as seen from these three- and six-sided examples.

- Distinctive reticulate seed coats of Primulaceae not preserved limiting level of identification.

- Dimensions (c.1mm across base) and morphology of these specimens suggest most likely taxa are *Anagallis arvensis*, *Lysimachia* spp., although may be some overlap with small seeds of *Primula* spp.

Examples of archaeological sites

Late Bronze Age midden at Potterne [WT1]; Iron Age ditch and pits at Battlesbury Bowl [WT2].

Modern ecology

Annuals and perennials. Habitats for the most common native and archaeophyte species range from cultivated soils of gardens and arable fields, shaded, often damp, hedgerows, woodland edges and rough grasslands.

Interpretative value

Interpretative value limited because wide habitat ranges. Most likely to be arable weed *Anagallis arvensis* if present as more than an occasional seed, but could be *Lysimachia* sp. if wet/damp areas in locality. *Anagallis arvensis* used to treat mental health problems and epilepsy, dropsy and other disorders (Web 1). Entire plant apart from roots used in extracts, tinctures, dried herb or teas, but poisonous in excess because of toxic saponins and cucurbitacins (Web 1).

Family: RUBIACEAE
Latin name: *Sherardia arvensis* L.
Common name: field madder
Anatomical element: seed

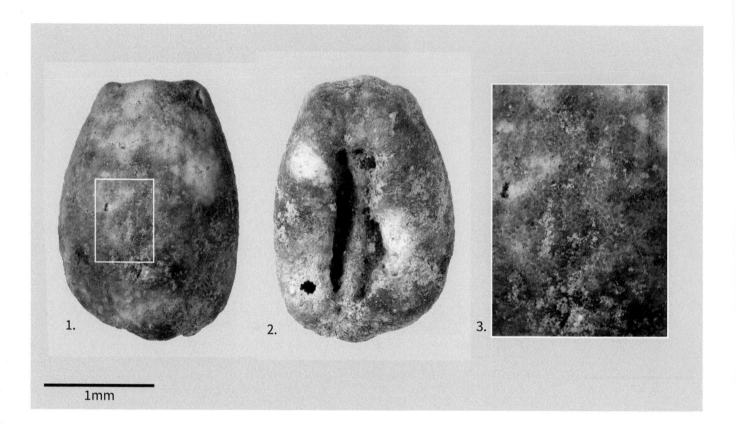

1mm

Image description

1-2. Dorsal and ventral views of seed, pericarp not preserved [B146].

3. High magnification image of dorsal surface.

Key diagnostic features and separation from similar taxa

- Distinctive blunt-end, oblong seed with ventral groove, septum along groove not always preserved. Small polygonal cells present on surface of seed.

- Similar to some Apiaceae eg *Conium maculatum* though with squared apical end and clear differences in cell patterns where pericarp not preserved. If preserved, *Sherardia* pericarp has horizontally elongated cells in ranks, rather than ladder-like cells of *Conium*.

Examples of archaeological sites

Iron Age ditch and pits at Battlesbury Bowl [WT2].

Modern ecology

Dry, open grasslands, arable fields, waste ground, waysides, sand dunes. More often found on neutral or basic soils.

Interpretative value

Indicator of dry, open, low nitrogen soils but requires moist, nutrient-rich conditions for mineralisation to take place. May indicate sudden smothering of vegetation with large amounts of midden-type waste or faecal material. Alternatively, derives from gathered vegetation brought to the site, or dung from livestock grazing on dry pastures. Fleshy roots can be used to make a pink/red dye (Web 1).

Family: RUBIACEAE
Latin name: *Galium aparine* **L.**
Common name: cleavers
Anatomical element: seed

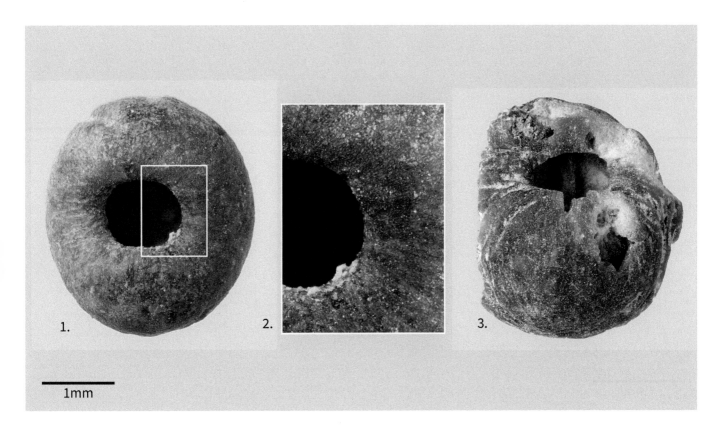

1mm

Image description

1. Example 1; spherical seed with large rounded aperture [B111].
2. High magnification image of example 1 showing rows of small angular cells radiating from aperture.
3. Example 2; incomplete seed angled to show cavity [B112]

Key diagnostic features and separation from similar taxa

- Bristly, thin pericarp not preserved.
- Large globose seeds have small sub-circular apertures with thick incurved rim.
- *G. aparine* fruits can vary in size. Large seeds can be identified to species level when whole, or where enough of aperture is preserved to show size and thickness of rim.

Examples of archaeological sites

Late Bronze Age midden at Potterne [WT1]; Iron Age pits at Battlesbury Bowl [WT2]; late Saxon cesspits at Northgate House, Winchester [HA1].

Modern ecology

Native annual of cultivated land, hedgerows, waste ground, waysides, showing preference for nutrient-rich soils.

Interpretative value

Being a nitrophilous plant, very likely to be found growing on or by middens and cesspits. May also be gathered with vegetation used as toilet wipes or for flooring etc. Hooked 'sticky' fruits likely to be dropped into latrines from clothing. May also have been a contaminant of foods. Grieve (1992) lists a wide range of medicinal uses of the whole plant, including use as diuretic, tonic, to purify the blood, to cure scurvy and skin complaints. Seeds can be roasted and used as a coffee substitute and dried plant can be used to make a soothing tea that was a rural remedy for colds.

Family: BORAGINACEAE
Latin name: *Lithospermum arvense* L.
Common name: field gromwell
Anatomical element: fruit and seed

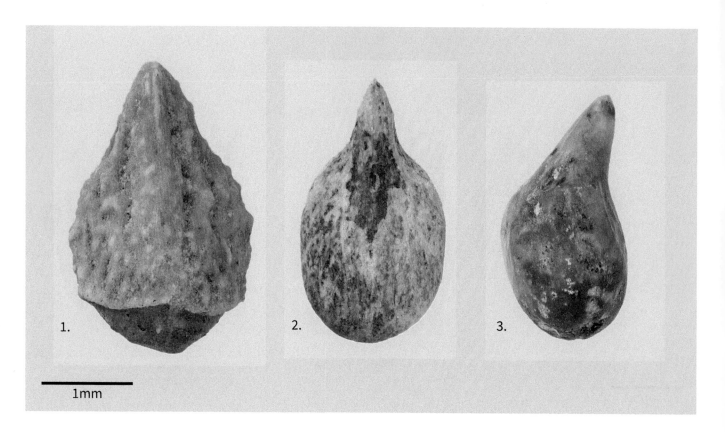

1mm

Image description

1. Example 1; whole fruit, dorsal view [B31].
2. Example 2; ventral view of seed, outer layers of pericarp not preserved though part of thin vascular mesophyll present [B15].
3. Example 3; seed without vascular mesophyll, side view showing curved profile [B51].

Key diagnostic features and separation from similar taxa

- Warty, thick-coated whole nutlet closely comparable with modern reference material.

- Seed has distinctive curved flask-shaped form.

- Whole nutlet and seed distinctive to species level. Seed differs from *L. officinale* which is more gradually tapered, ovate rather than flask-shaped.

Examples of archaeological sites

Late Bronze Age midden at Potterne [WT1]; Iron Age ditch, pits and post-holes at Battlesbury Bowl [WT2]; early Iron Age ditch at Flint Farm, Danebury Environs Project [HA3]; medieval and post-medieval pits from Leicester Shires [LR3].

Modern ecology

Archaeophyte, annual of arable, disturbed ground and open grassy places on light, dry, often calcareous soils. Seed is short-lived so populations depend on regular soil disturbance for survival (Web 2).

Interpretative value

Often present in midden-type deposits on calcareous soils. May indicate deposition of cereal processing waste or plants weeded from fields. Plants could also grow in and around features such as ditches in disturbed areas.

Family: BORAGINACEAE
Latin name: *Myosotis* **sp.**
Common name: forget-me-not
Anatomical element: seed

1mm

Image description

1. Example 1; fruit, dorsal and ventral views, with at least some layers of pericarp preserved, basal attachment scar obscured by broken pericarp [B23].

2. Example 2; fruit, ventral view, basal attachment scar visible. High magnification image showing sinuous cell pattern of pericarp [B60].

3. Example 3; seed, ventral view, with small area of pericarp preserved at apex [B169].

Key diagnostic features and separation from similar taxa

- Ovate shaped fruit with apex pinched on ventral side and slight rim around edges. Faint, slightly sinuous, irregular cell pattern giving shiny appearance to surface.

- Closely comparable with modern reference specimens where pericarp preserved. Similar morphology where no pericarp, but no ridges around rim, depression in embryo/endosperm in position of basal scar and asymmetrical polygonal cell pattern.

- Fruits of Myosotis difficult to identify species level due to similarities in size and form, though some variation in outlines seen in these examples.

Examples of archaeological sites

Late Bronze Age midden at Potterne [WT1]; Iron Age ditch and pits at Battlesbury Bowl [WT2]; Anglo-Saxon pits at Abbots Worthy [HA7].

Modern ecology

Seven common or locally common native and archaeophyte British species found in a wide range of habitats and soil types, from weed of open, well drained cultivated and disturbed soils to woods, bogs, marshes, rock ledges and damp grasslands. Annuals and perennials included in group.

Interpretative value

Difficulty in identifying to species level results in wide range of habitat preferences, limiting interpretative value. According to modern records, most common weed species is *M. arvensis*, likely to be present as contaminant of cereal-based foods or amongst grassy material used as toilet wipes.

Family: SOLANACEAE
Latin name: *Hyoscyamus niger* **L.**
Common name: henbane
Anatomical element: seed

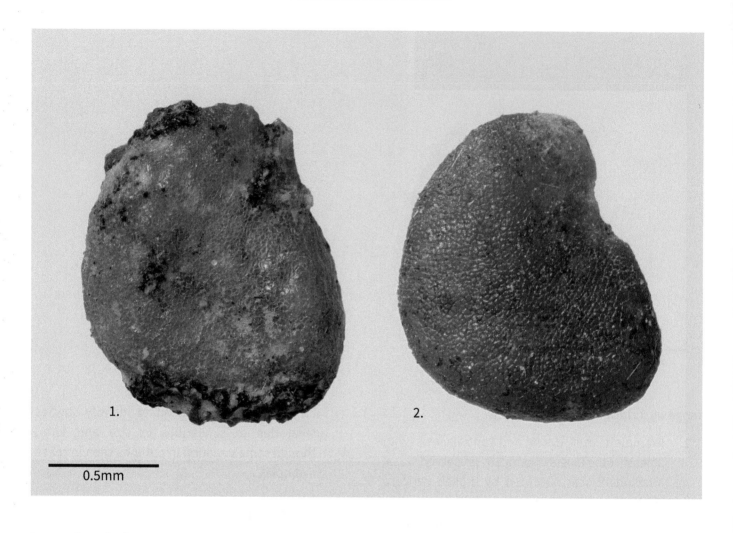

1.

2.

0.5mm

Image description

1. Example 1; seed with traces of black undulate outer epidermal layer of seed coat surviving [B7].

2. Example 2; seed with seed coat not preserved [B67].

Key diagnostic features and separation from similar taxa

- Sub-triangular shape, thicker at the base, thinning towards the micropylar end (upper right).

- Most commonly found without seed coat.

- Gross morphology is the most distinctive factor making henbane seeds easy to separate from other Solanaceae. *Solanum nigrum* (see page 37) and *S. dulcamara* more evenly flattened.

Examples of archaeological sites

Iron Age ditches and pits at Battlesbury Bowl [WT2]; Anglo-Saxon cesspit at Abbots Worthy [HA7]; post-medieval cesspit at Leicester Shires [LR3].

Modern ecology

Annual and biennial archaeophyte of dry, calcareous and sandy soils. Prefers disturbed, nitrogen-rich soils, typically found in farmyards and on middens.

Interpretative value

Indicator of nutrient-enriched soils, so likely to have grown on middens and near cesspits, though could be some medicinal use. Although poisonous, henbane cultivated for medicinal use in past, seeds being particularly valued for complaints such as toothache and rheumatism (Grieve 1992).

Family: SOLANACEAE
Latin name: *Solanum nigrum* **L.**
Common name: black nightshade
Anatomical element: seed

1mm

Image description

1. Example 1; some of reticulate seed coat preserved, perhaps partly waterlogged [B68].

2. Example 2; more rounded form and more acute apical end demonstrating variations in form [B70].

3. Example 3; face view and profile [B69].

Key diagnostic features and separation from similar taxa

- Flattened oval shape with indentation at hilar end and clear polygonal cell pattern beneath seed coat.

- Seed of *S. dulcamara* is larger and more rounded but with similar clear cell pattern, also having some variations in form.

- Some other species present in the British Isles may be similar in size and form but are introduced and rare, mostly originating from the New World.

Examples of archaeological sites

Medieval cesspit at land near Ferrer's Lane, Huntingdon [CB1]; medieval cesspits at Bradwell's Court, Cambridge [CB2].

Modern ecology

Annual weed of cultivated and waste ground, often found on nutrient-enriched soils. Can also grow in damp shady places along waysides.

Interpretative value

Likely to have grown close to features such as middens and cesspits due to preference for nutrient-rich soils. Also used medicinally in the past which could account for presence in cesspits. Although poisonous and avoided by livestock, concentration of active principle, solanine, varies with season and said not to harm adults when ripe berries consumed (Grieve 1992, 583). Some decoctions, made from green parts of the plant, were formerly used to treat asthma, rheumatism, bronchitis and skin complaints (Stuart 1987).

Family: LAMIACEAE
Latin name: *Lamium/Ballota/Marrubium* - type
Common name: dead-nettle/black horehound/white horehound-type
Anatomical element: seed

1mm

Image description

1. Example 1; dorsal and ventral views, good preservation of attachment scar end, possible mineralisation of some of pericarp forming flanges more typical of *Ballota nigra* [B57]

2. Example 2; shorter seed with more squared upper end, more typical of *Marrubium vulgare* and some *Lamium* species [B171].

Key diagnostic features and separation from similar taxa

- Oblong, fairly parallel-sided seed with triangular cross-section, rounded or slightly squared upper end. Pericarp not often preserved.

- Several Lamiaceae with this morphology and size range when pericarp is removed (eg *Glechoma hederacea*), so closest matches and archaeobotanically most commonly found taxa used to form 'type'.

- *Prunella vulgaris* (page 40) can be separated by flatter, wider cross-section, no ridge on ventral surface.

Examples of archaeological sites

Late Bronze Age midden at Potterne [WT1]; medieval cesspit at Causeway Lane, Leicester [LR2]; late medieval and post-medieval pits at Bonner's Lane, Leicester [LR4].

Modern ecology

Annual species in group are archaeophytes of open cultivated and waste ground. Perennial species grow in grasslands, hedgerows and rough ground. Some species show preference for nitrogen-rich soils eg *Marrubium vulgare* and *Lamium album* (Hill *et al* 1999).

Interpretative value

Some species likely to have been growing on nutrient-rich soils by middens and cesspits. Taxa found in grassland habitats probably present amongst vegetation used as toilet wipes. Wide range of medicinal uses for shoots and leaves from plants of this group including reduction of blood loss (*Lamium* spp.), for respiratory complaints and as vermifuge (*Ballota nigra*) and as tonic and relief of coughs and colds (*Marrubium vulgare*) (Web 1).

Family: LAMIACEAE
Latin name: *Galeopsis* sp.
Common name: hemp-nettle
Anatomical element: fruit

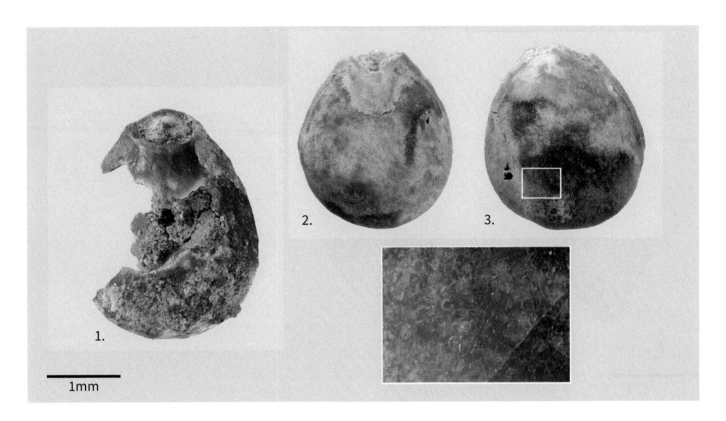

1mm

Image description

1. Example 1; partial preservation of fruit, ventral view showing intact attachment scar [B133].

2. Example 2; fruit, ventral view. Outer layers of pericarp including verrucose ornamentation not preserved [B134].

3. Example 3; dorsal view of fruit and high magnification image showing faint pattern of dense rounded/polygonal cells [B134].

Key diagnostic features and separation from similar taxa

- Broadly ovate fruit with large, distinctive circular attachment scar and smooth surface (verrucose ornamentation of outer pericarp layers not preserved). Morphologically, fruits closely resemble modern reference material where features of pericarp preserved.

- Only two of five *Galeopsis* species found in British Isles are common today. Examples above closely resemble *G. tetralix* though attachment scars and fruits of *G. bifida* and *G. speciosa* have very similar dimensions.

- Most similar genus in Lamiaceae, *Stachys* sp., has fruits that are either smaller, narrower and/or with smaller attachment scars.

Examples of archaeological sites

Late Bronze Age midden at Potterne [WT1]; Late Bronze Age midden at East Chisenbury [HA2].

Modern ecology

All British native or archaeophyte species grow as annuals in a range of moist, fairly shaded habitats such as ditches, fens, river banks, waysides, but also as arable and waste ground weeds.

Interpretative value

Both examples recovered from Late Bronze Age middens, perhaps having been smothered by dumped organic waste, or discarded amongst crop processing waste or plants weeded from fields.

Family: LAMIACEAE
Latin name: *Prunella vulgaris* **L.**
Common name: selfheal
Anatomical element: seed

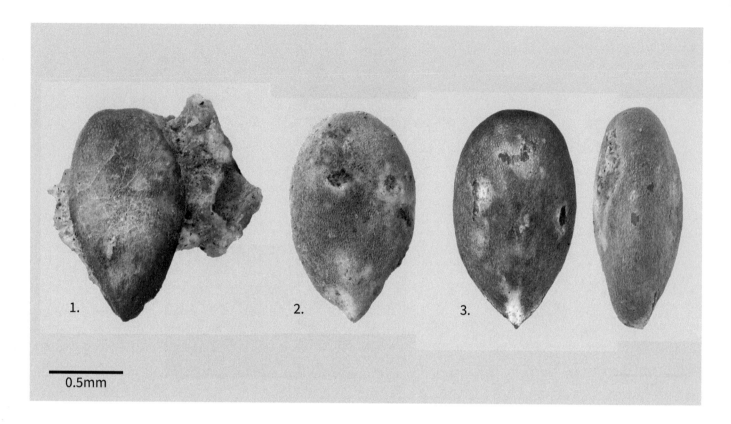

0.5mm

Image description

1. Example 1; dorsal view of seed (embedded in concretion), outer layers of pericarp not preserved, possible fungal hyphae present [B114].

2. Example 2; ventral view, outer layers of pericarp not preserved, slightly sinuate-walled cell pattern [B115].

3. Example 3; ventral and side view, pericarp not preserved [B113].

Key diagnostic features and separation from similar taxa

- Oval, dorso-ventrally compressed seed with slightly flattened area down central surface (corresponding to ornamentation on intact fruit), most obvious on ventral side. Sometimes slightly indented upper end as in example 1.

- Separated from *Lamium*-type (cf. page 38) primarily by flattened form and lack of ventral ridge.

Examples of archaeological sites

Iron Age ditch and pits at Battlesbury Bowl [WT2]; post-medieval pit at Bonner's Lane, Leicester [LR4].

Modern ecology

Native patch-forming perennial of a range of grassland habitats, including meadows, pastures, waysides, waste ground. Widespread and common.

Interpretative value

Could be present amongst grassy vegetation used as toilet wipes, in dung from livestock feeding on wet pastures or growing near to middens and cesspits, since often associated with moist fertile soils. Wide range of medicinal uses so may have been consumed. Taken internally as tea for diarrhoea, fevers, mouth complaints and externally for its antibacterial action against infections. Can be used fresh or dried (Web 1).

Family: OROBANCHACEAE
Latin name: *Rhinanthus* sp.
Common name: yellow-rattle
Anatomical element: seed

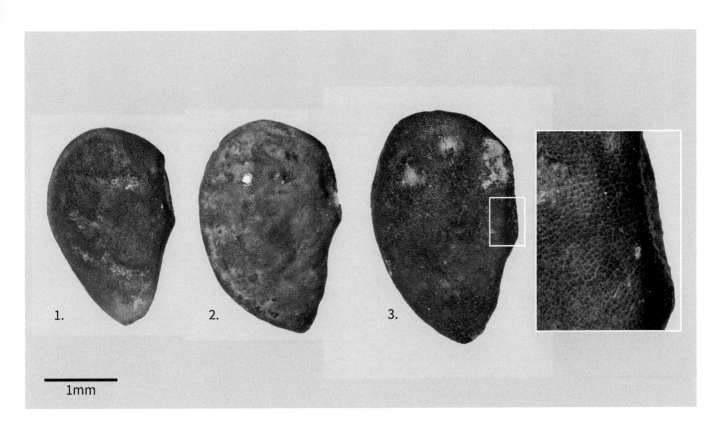

Image description

1-2. Examples 1 & 2; seeds with seed coats and marginal wings not preserved [B55] [B97].

3. Example 3; seed (no seed coat) with high magnification insert showing typical polygonal cell pattern of endosperm/embryo [B98].

Key diagnostic features and separation from similar taxa

- Oval, flattened seeds with clear pattern of polygonal cells evenly spread across the seed.

- Similar to *Ranunculus* sp. (page 4) in general morphology but cell patterns differ, *Rhinanthus* having larger, squarer, slightly convex cells. Distinctions between *Rhinanthus* species not investigated due to scarcity of reference material for *R. angustifolius*.".

Examples of archaeological sites

Late Saxon cesspit, Winchester Northgate House [HA1]; 13th/14th century garderobe, Jennings Yard, Windsor [BK1].

Modern ecology

The two British species of yellow-rattle are annual hemi-parasites of grasses. *R. angustifolius* once a frequent weed of arable in eastern Britain, now rare. *R. minor* is the only common and widespread species today.

Interpretative value

R. minor could be preserved amongst hay used as toilet wipes, stable waste, flooring or to soak up fluids. *R. angustifolius* may have been consumed as a contaminant of cereal-based foods or discarded amongst processing waste in the past.

Family: ASTERACEAE
Latin name: *Cirsium/ Carduus*-type
Common name: thistle-type
Anatomical element: seed

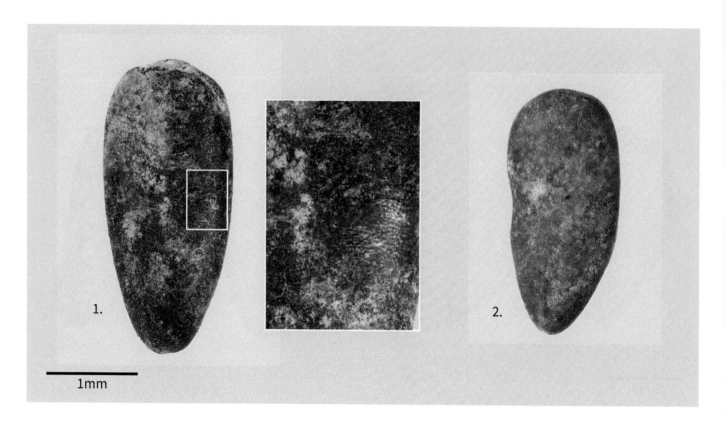

1mm

Image description

1. Example 1; large seed, pericarp of achene not preserved, plus high magnification image showing cell pattern [B152].

2. Example 2; small seed, pericarp of achene not preserved [B153].

Key diagnostic features and separation from similar taxa

- Obovate seed with transversely elliptic cross section and indistinct, small polygonal cell pattern.

- Some variation in size, these examples roughly comparable with *Cirsium vulgare* (c. 4mm length) and *Cirsium arvense* (c. 2.5mm).

- Similar shaped Dipsacaceae fruits differ in seed morphology, most seeds having pointed rather than rounded apical ends.

- May be overlaps in size with some other Asteraceae genera, hence '-type'.

Examples of archaeological sites

Iron Age ditch and pits at Battlesbury Bowl [WT2]; early medieval pit at Bonner's Lane, Leicester [LR4].

Modern ecology

Annuals, bennials and perennials mainly found in grasslands, hedgerows, waysides and waste places but also arable weeds (eg *Cirsium arvense*). A few species mainly found in wet to damp places.

Interpretative value

Achenes may have been shed from plants growing close to middens and cesspits, been present in animal dung, or as contaminants of crop plants. Inability to identify to genus or species level limits interpretative value.

Family: ASTERACEAE
Latin name: *Centaurea cyanus* **L.**
Common name: cornflower
Anatomical element: fruit and seed

1mm

Image description

1. Example 1; fruit, pericarp incomplete, plus high magnification image of pericarp showing vertically aligned chains of narrow cells [B151].

2. Example 2; seed, pericarp/seed coat not preserved, plus high magnification image of seed surface showing smooth surface layer of irregular polygonal cells [B150].

Key diagnostic features and separation from similar taxa

- Most often found as seed, without pericarp preserved .

- Key identification factor distinguishing from other *Centaurea* species is length of attachment scar (seen in mineralised seed as sloping lower section of right side of seed) = 1/3 or more of achene length (Knörzer in Greig 1991, 97). Other species of *Centaurea* have shorter angled side.

- Cf. *Carduus/Cirsium*–type (page 42), similar in size and form but lacking large attachment scar resulting in rounded base rather than pointed and sharply angled.

Examples of archaeological sites

Medieval cesspit at Chesil Street, Winchester [HA6]; 13th/14th century garderobe at Jennings Yard, Windsor [BK1].

Modern ecology

Annual archaeophyte, once a frequent arable weed, now mainly found on waste ground and roadsides. Archaeobotanical records suggest preference for light sandy soils (Greig 1991, 106).

Interpretative value

Greig (1991, 102) notes cornflower (waterlogged and mineralised) is common component of medieval faecal deposits. Probably represent contaminants of cereal-based foods. Flowers have been used as tonic and stimulant (Grieve 1992).

Family: ASTERACEAE
Latin name: *Lapsana communis* L.
Common name: nipplewort
Anatomical element: fruit and seed

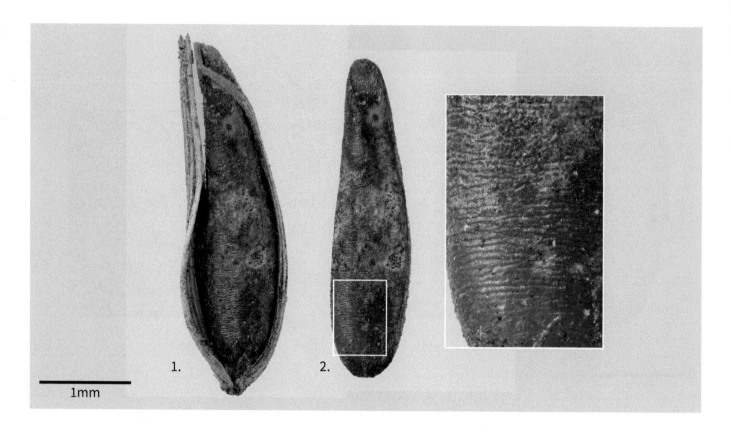

1mm

Image description

1. Fruit with waterlogged pericarp dried out and split open showing mineralised seed [B21].

2. Mineralised seed removed from pericarp with high magnification insert showing cell detail [B21].

Key diagnostic features and separation from similar taxa

*   Seed distinctive in its elongated ovoid shape. Embryo not preserved in this example, leaving apical depression which may or may not be typical. Horizontal ridged cell pattern of endosperm/embryo clearest in imbibed reference material.

*   This specimen has confirmed identification because of presence of waterlogged pericarp. Where pericarp not present it may be difficult to separate from other c. 3 to 4mm long members of the Asteraceae. However, examination of *Hypochaeris* sp. and *Leonotodon* sp. produced more slender seeds not broadening significantly towards base.

Examples of archaeological sites

Roman cesspit at Bedminster, Bristol [BR1]; base of 13th to 16th century London City ditch, Aldersgate [LD2].

Modern ecology

Common annual to perennial herb of disturbed and shady places on wide range of soil types. Can grow as weed of arable crops.

Interpretative value

Called nipplewort because of belief that it could be used to treat ulcerated nipples (Grigson 1987). Likely to have been blown into ditch or growing nearby in this example. Could also have been contaminant of food, present in vegetation used as toilet wipe or flooring where recovered from cesspit.

Family: ASTERACEAE
Latin name: *Anthemis/Glebionis/Tripleurospermum* - type
Common name: chamomile/marigold/mayweed-type
Anatomical element: seed

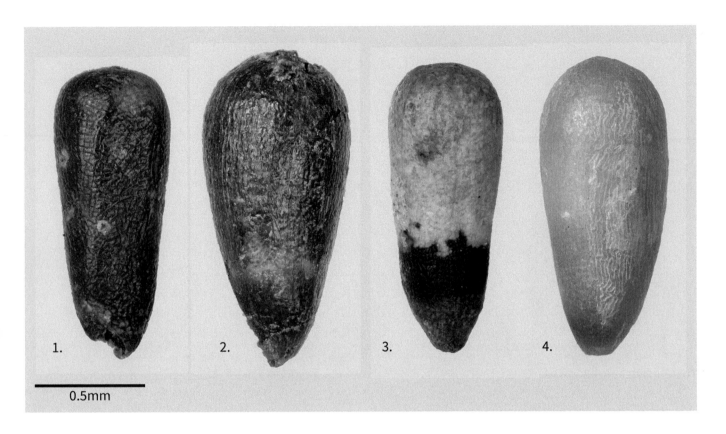

0.5mm

Image description

1-4. Four examples of seeds, pericarps not preserved though traces at both ends of example 2 and partial survival of vertically elongated, sinuate-walled cells on surface of example 4 [B62] [B172] [B173] [B174].

Key diagnostic features and separation from similar taxa

- Small (c. 1.25mm long) obovate seeds with circular cross-section, rounded upper end gradually narrowing towards base, coarse pattern of polygonal, thick-walled cells.

- Several common taxa frequently recovered from archaeobotanical samples fall into this size and morphological group, most notably *Anthemis cotula*, *Glebionis segetum* and *Tripleurospermum inodorum*. Also likely to be overlaps with other, less common species of Asteraceae, hence use of '-type'.

Examples of archaeological sites

Medieval secondary fill of cesspit at Trowbridge [WT4]; *Anthemis cotula* and *Anthemis* sp. in post-medieval pits at St Peter's Lane, The Shires, Leicester [LR3].

Modern ecology

Cited species all annual archaeophytes of arable and waste ground. *A.cotula* shows preference for heavy soils, *G.segetum* for lighter, calcium deficient soils, *T. inodorum* for fertile soils.

Interpretative value

Most likely present as crop contaminants or amongst vegetation used as toilet wipes, flooring etc.

Family: VALERIANACEAE
Latin name: *Valerianella* sp.
Common name: cornsalad
Anatomical element: seed

0.5mm

Image description

1. Example 1; dorsal, side and ventral views, pericarp not preserved [B59].

2. Example 2; dorsal view, pericarp not preserved [B101].

Key diagnostic features and separation from similar taxa

- Slender ovate seed with curved profile and distinctive sinuous-walled reticulate cell pattern.

- Curved profile and pointed apex distinguishes this genus from similar sized ovate seeds such as biconvex *Urtica urens*.

- Shiny surface and sinuous cell pattern may help to identify incomplete seeds.

- Similarities in morphology and cell patterns amongst British species of *Valerianella* with pericarp removed mean identification to species level is unlikely.

Examples of archaeological sites

Late Bronze Age midden at Potterne [WT1]; Iron Age midden deposits in pits and ditches at Battlesbury Bowl [WT2].

Modern ecology

All British species are annuals, most grow on arable land and wide range of other disturbed habitats as well as open grasslands. Prefer open habitats on fairly infertile, well-drained soils (Ellenberg 1988).

Interpretative value

V. locusta grown as salad vegetable today, so at least some species likely to have been consumed in the past, perhaps including fruiting parts. From examples given, recovery of *Valerianella* sp. from two midden-type sites on free-draining soils suggests seeds were either present as arable weeds amongst processing waste or growing in disturbed areas smothered by midden deposits.

Family: APIACEAE
Latin name: *Scandix pecten-veneris* **L.**
Common name: shepherd's needle
Anatomical element: fruit/seed

1mm

Image description

1. Example 1; dorsal and ventral views of incomplete ridged mericarp, long dorsally flattened beak not preserved [B77].

2. Example 2; dorsal and ventral views of incomplete mericarp, most of pericarp and beak not preserved [P78].

Key diagnostic features and separation from similar taxa

- Long, slender mericarp distinctive in size and morphology.
- Larger and more slender and parallel sided than other long Apiaceae, eg *Chaerophyllum* sp.

Examples of archaeological sites

Medieval cesspit at Chesil Street, Winchester [HA6]; medieval cesspit at Ferrar's Road, Huntingdon [CB1].

Modern ecology

Archaeophyte, annual of arable fields, particularly on calcareous soils. Occasionally found on waysides close to abandoned arable land. Less common today than formerly.

Interpretative value

Likely contaminant of cereal-based foods or present amongst vegetation used as toilet wipes, flooring etc.

Family: APIACEAE
Latin name: *Coriandrum sativum* L.
Common name: coriander
Anatomical element: fruit and seed

1mm

Image description

1. Example 1; Fragment of mericarp showing pericarp with distinctive wavy lines between ridges [B136].

2. Example 2; One of two hemispherical seeds from within mericarp, dorsal & ventral view [B137].

Key diagnostic features and separation from similar taxa

- Globose mericarps (pair not separating at maturity) with characteristic sinuous ornamentation, distinctive enough for fragments to be confidently identified.

- Where individual hemispherical seeds preserved their morphology and cell patterns are distinctive.

Examples of archaeological sites

Roman cesspits at Silchester [HA2]; middle Saxon cesspits at St. Mary's Stadium, Southampton [HA5].

Modern ecology

Annual grown for aromatic leaves and seeds, possibly originating from S.W Asia and Armenia. One of earliest crops of Old World grown as condiment (Zohary *et al* 2013, 163).

Interpretative value

Only rarely grows as casual in British Isles (Web 2) so presence is good indicator of food or faecal waste. Primarily found in urban cesspits dating from Roman period onwards (van der Veen *et al* 2008). Essential oils in the plant thought to have antibiotic properties. Long history of use to treat mouth ulcers, digestive and gastric complaints etc. (Web1).

Family: APIACEAE
Latin name: *Aethusa cynapium* L.
Common name: fool's parsley
Anatomical element: seed

1mm

Image description

1. Example 1; dorsal view of seed (pericarp not preserved) [B34].

2. Example 2; ventral view; smaller seed showing size variation [B166].

3. Example 3; dorsal, side and ventral views [B76].

Key diagnostic features and separation from similar taxa

- Sculpted pericarp not preserved, slight ridges usually visible on dorsal surface.

- Ventral side flat with low ridge at apical end.

- Similar oval outline to Apiaceae such as *Conium maculatum* but more flattened dorso-ventrally. Lack of ventral groove is distinctive.

Examples of archaeological sites

Late Bronze Age midden, Potterne [WT1]; Iron Age pit, Battlesbury Bowl [WT2]; medieval pit from Huntingdon, Cambs [CB1].

Modern ecology

Widespread annual herb of cultivated land, waste ground, hedgebanks. Also grows as arable weed, mainly on dry soils.

Interpretative value

Grieve (1992) notes that, although poisonous, fool's parsley has been used medicinally in past for gastro-intestinal complaints such as diarrhoea and gastroenteritis. May be present as a contaminant of cereal-based foods in cesspits. Due to preference for drier soils with average nitrogen levels unlikely to have been growing around middens and cesspits.

Family: APIACEAE
Latin name: *Anethum graveolens* L.
Common name: dill
Anatomical element: seed

1mm

Image description

1. Example 1; dorsal, ventral and side views, pericarp not preserved [P48]
2. Example 2; dorsal and ventral views, pericarp not preserved [P155].

Key diagnostic features and separation from similar taxa

- Dorsally compressed fruits, slightly angular dorsal side though prominent dorsal and lateral ridges of pericarp lost.
- Length/breadth ratio smaller than *Carum carvi* and *Cuminum* spp. which are longer and narrower.
- Similar to *Foeniculum vulgare* (see page 51) but more dorsally compressed (flatter cross-section), curved outline in dorsal and ventral views, low dorsal ridges rather than undulations.

Examples of archaeological sites

Middle Saxon cesspits at St. Mary's Stadium, Southampton [HA5]; post-medieval cesspits at The Shires, Leicester [LR3].

Modern ecology

Possible native of Mediterranean basin and West Asia with long history of use in Middle East since fourth millennium BP (Zohary *et al* 2013), imported to British Isles from Roman period (van der Veen *et al* 2008).

Interpretative value

Indicator of faecal waste and cesspits (Smith 2013). Aromatic fruits have many medicinal and culinary uses; speed-up healing, aids digestion, alleviates flatulence, dyspepsia, constipation, diarrhoea; used as a pickling spice and flavouring foods (Web 1; Grieve 1992).

Family: APIACEAE
Latin name: *Foeniculum vulgare* L.
Common name: fennel
Anatomical element: seed

1mm

Image description

1. Example 1; dorsal, ventral and side views, most of pericarp not preserved including ribs, remains of stylopodium present on apex [B160].

2. Example 2; dorsal and ventral views, some pericarp/seed coat preserved, with high magnification image of cell pattern [B159].

Key diagnostic features and separation from similar taxa

- Ovoid/oblong seed, roughly square in cross section with undulating ridges. If present, stylopodium on apical end relatively tall.

- Similar to dill (see page 50) but not dorsally flattened, more prominent undulating dorsal and lateral ridges, more parallel sided.

- Both dill and fennel seeds quite variable in length.

Examples of archaeological sites

Middle Saxon cesspits at St. Mary's Stadium, Southampton [HA5]; post-medieval cesspit at Cockspur Street, London [LD3].

Modern ecology

Tall perennial herb, probably originating in southern Europe and Mediterranean (Stuart 1987), now considered archaeophyte. Introduced to British Isles in Roman period (van der Veen *et al* 2008). Naturalised on wasteland and well-drained soils in sunny locations (Stuart 1987).

Interpretative value

Indicator of cesspits and human faecal deposits (Smith 2013). Wide range of medicinal and culinary uses, particularly seeds which can be chewed or made into tea to aid digestive disorders, respiratory congestion, reduce fevers and reduce colic in babies (Grieve 1992).

Family: APIACEAE
Latin name: *Conium maculatum* **L.**
Common name: hemlock
Anatomical element: seed

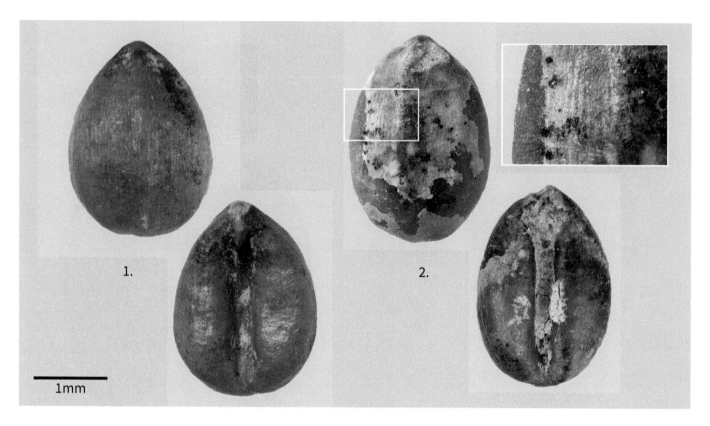

1.

2.

1mm

Image description

1. Example 1; dorsal and ventral views of seed (highly imbibed), pericarp not preserved [B5].

2. Example 2; dorsal and ventral views, dorsal ridges more distinct, probably less imbibed, traces of pericarp/seed coat preserved. High magnification image showing partial preservation of ladder-like cell pattern [B144].

Key diagnostic features and separation from similar taxa

- Identification confirmed for Example 1 as waterlogged pericarp with distinctive undulate-crenulate ridges present when recovered, but removed for photograph.

- Ovoid seeds, slightly laterally compressed with a wide furrow.

- Distinctive ladder-like cell pattern on surface where pericarp/seed coat present.

- Broader and plumper than similar-sized Apiaceae such as *Pimpinella* sp., cell pattern distinctive.

Examples of archaeological sites

Base of 13th to 16th century London City Ditch at Aldersgate [LD2]; early post-medieval cesspit, Bonners Lane, Leicester [LR4].

Modern ecology

Biennial archaeophyte, damp places on roadsides, waste ground, stream banks and ditches. Favours nutrient-rich soils. Widespread and common.

Interpretative value

Indicator of damp, nutrient-rich places so likely to have been growing in the vicinity of cesspits and middens. Although poisonous, fruits have been used medicinally both externally and internally as antispasmodic and sedative (Grieve 1992). Ointments, tinctures and inhalations used for conditions ranging from epilepsy to skin and joint complaints (Grieve ibid).

Family: APIACEAE
Latin name: *Bupleurum rotundifolium*-type
Common name: thorow-wax -type
Anatomical element: seed

1mm

Image description

1. Example 1; dorsal and ventral views, pericarp not preserved [B43].

2. Example 2; dorsal and ventral views, pericarp not preserved [B64].

Key diagnostic features and separation from similar taxa

- Small pinched stylopodium at apex, parallel sided oblong seed, laterally compressed with roughly squared cross-section, c. 3-3.5mm long.

- Pericarp not preserved in these examples. Low rounded ridges on dorsal side.

- Possible slight overlap in size and morphology with some other species of *Bulpeurum* and other members of Apiaceae (eg *Petroselenium* sp.) though today most are rare or introduced to British Isles.

- Most other similar sized Apiaceae are less compressed and parallel sided.

Examples of archaeological sites

Saxo-Norman cesspit, Wllington to Steppingley pipeline [BD1]; Early medieval cesspit, Trowbridge, Wilts [WT4].

Modern ecology

Archaeophyte, formerly an arable weed of well-drained chalk and limestone soils, now rare.

Interpretative value

Presence in cesspits most likely to be as a contaminant of cereal-based foods or discarded processing waste. However, *B. falcatum* has been used in traditional Chinese medicine since the 1st century BC for liver and stomach complaints (Web 3) so medicinal use of *B. rotundifolium* in the past is possible.

Family: APIACEAE
Latin name: *Torilis* sp.
Common name: hedge-parsley
Anatomical element: fruit and seed

1.

2.

1mm

Image description

1. Example 1; dorsal and ventral views with patchy preservation of pericarp, including areas of tuberculate exocarp (possibly bases of spines) and ridges [B33].

2. Example 2; dorsal and ventral views, appears to be tuberculate fruit of *T. nodosa* though broken/ incompletely preserved spines may have similar appearance [B142].

Key diagnostic features and separation from similar taxa

- Where spines or tubercles preserved may be possible to identify to species level. However, spine bases may look similar to tubercles.

- Heavily ornamented exocarp not always preserved so gross morphology and length/ breadth measurements important in separating from similar taxa.

Examples of archaeological sites

Late Bronze Age midden at Potterne [WT1]; early Iron Age ditch at Flint Farm, Danebury Environs Project [HA3]; medieval cesspit at Jennings Yard [BK1].

Modern ecology

Torilis nodosa and *T. arvensis* are weeds of arable, now only local. *T. japonica* more widespread, found in grassy places and hedgerows, often on nutrient-rich soils.

Interpretative value

Torilis sp. may be either food contaminant or present in grassy vegetation used as toilet wipes, flooring etc. *T. japonica* could have grown close to middens and cesspits.

Family: APIACEAE
Latin name: *Daucus carota* L.
Common name: carrot
Anatomical element: fruit/seed

1mm

Image description

1. Example 1; dorsal and ventral views, two orange vittae preserved. High magnification image showing ladder-like cell pattern of lower pericarp/seed coat layers [B14].

2. Example 2; dorsal and side views showing compressed profile [B66].

Key diagnostic features and separation from similar taxa

- Oblong, dorsally compressed mericarp with wide furrow.

- Distinctive ladder-like cell pattern visible if lower layer of pericarp/seed coat preserved.

- Ridges of pericarp with spines and cilia usually not preserved but vittae sometimes present as in example 1, aiding identification (see Tutin 1980, 183).

- Other Apiaceae of this size and form are less dorsally compressed, eg *Torilis* sp. (page 54).

Examples of archaeological sites

Late Bronze Age midden at Potterne [WT1]; Anglo-Saxon cesspits at Abbots Worthy [HA7]; medieval cesspit at 70, Stour Street, Canterbury [KT1].

Modern ecology

Biennial plant often found on poor soils, particularly grasslands on calcareous soils. Includes chalk downlands, rough grassland, waysides and waste places.

Interpretative value

Relatively frequently mineralised because common on calcareous soils. May be preserved by being smothered under midden material. In cesspits, may have been used for flavouring. Bruised seeds used medicinally for wide range of complaints including jaundice, colic, dysentery (Grieve 1992). Could also be gathered amongst grassy vegetation for use as toilet wipes, floor covering etc.

Family: CYPERACEAE
Latin name: *Eleocharis* sp.
Common name: spike-rush
Anatomical element: fruit and seed

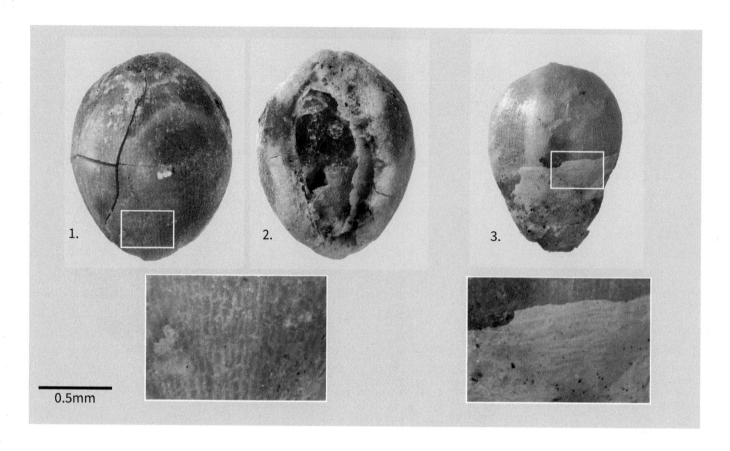

0.5mm

Image description

1. Example 1; whole achene slightly angled to show scar from loss of apical tubercle, plus high magnification image of sinuous-walled oblong cells of pericarp [B139].

2. Example 1; broken side of achene showing thick, spongy pericarp and cross-section of seed with white transverse cell layer [B139].

3. Example 2; seed with distinctive white layer of transversely-elongated cells, shown in high magnification insert [B140].

Key diagnostic features and separation from similar taxa

- Where whole fruit is preserved it closely resembles modern reference specimens. Seed less often preserved but recognisable by distinctive obovoid shape coming to slight point at apex, particularly where white horizontal cell layer preserved (characteristic of many Cyperaceae seeds).

- Distinctive morphology allowing identification to *Eleocharis* sp. though possibly no further.

Examples of archaeological sites

Late Bronze Age midden at Potterne [WT1]; Anglo-Saxon cesspits from Abbots Worthy [HA7]

Modern ecology

All British species are rhizomatous perennials found on range of wet to damp soils, habitats including waterside, marsh, fen, bog, dune-slack, ditches.

Interpretative value

Most likely a component of vegetation gathered as toilet wipes, flooring, roofing material or to dampen odours and soak up fluids. On middens may derive from dung of animals grazed on marshy ground, animal bedding or waste fodder. May also have been smothered by dumping of midden material in wetter areas.

Family: CYPERACEAE
Latin name: *Carex* - type
Common name: sedge-type
Anatomical element: seed

0.5mm

Image description

1. Example 1; short, trigonous seed with horizontally elongated cells on lower half, outer layers of pericarp not preserved [B37].

2. Example 2; small, lenticular seed with thin layer of horizontally elongated cells and some vertically elongated in central area, remainder of pericarp not preserved [B39].

3. Example 3; long trigonous seed with horizontally elongated cell layer, outer layers of pericarp not preserved [B38].

Key diagnostic features and separation from similar taxa

- Preservation of a layer of narrow horizontally-elongated cells surrounding seed is characteristic of the family and helps to separate from other trigonous seeds such as *Rumex* sp., as well as general morphology.

- Although *Carex* sp. is the most likely genus, with the loss of surface features of the fruit a few other members of Cyperaceae could be similar, hence the use of '-type'.

- Not likely to be identifiable to species level, although example 2 has more distinctive shape and could be tentatively identified as cf. *Carex echinata*.

Examples of archaeological sites

Late Bronze Age midden at Potterne [WT1]; early and late medieval cesspits and post-medieval drain at St Nicholas Place, Leicester [LR5].

Modern ecology

Sedges and other members of Cyperaceae primarily found in damp to wet habitats such as damp grassland, woodland, marshes and watersides.

Interpretative value

Commonly recovered from cesspits and middens. In faecal deposits may represent materials used as toilet wipes, for reducing smells/soaking up liquids and flooring. In middens could derive from animal bedding, dung, fodder, discarded flooring materials or local vegetation smothered by dumped material.

Family: POACEAE
Latin name: *Poa* - **type**
Common name: meadow grass-type
Anatomical element: seed

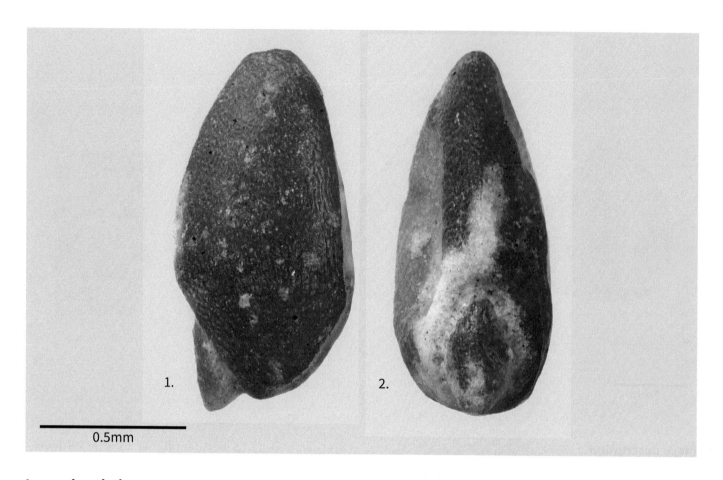

0.5mm

Image description

1. Seed, elongated cells of pericarp not preserved, side view [P116].

2. Seed, elongated cells of pericarp not preserved, dorsal view [P116]. This example typical of *Poa annua* L.

Key diagnostic features and separation from similar taxa

*   Small, wedge-shaped grain, wider at base with pointed apical end and prominent embryo. Ventral groove barely visible.

*   Possibility of size (c. 1mm long) and morphological overlaps with some similar small-seeded grasses, for example *Alopecurus* spp., *Phleum* spp., hence use of '-type'.

*   Several other grass seeds differing in size and shape have been preserved by mineralisation. *Poa*-type is one of more commonly found taxa.

Examples of archaeological sites

Late Bronze Age midden at Potterne [WT1]; Late Bronze Age midden at East Chisenbury [WT3].

Modern ecology

Poa annua is native annual or perennial, common and widespread on open and disturbed ground, including cultivated land, waste ground, waysides, gardens. Flowers all year round. Other *Poa* species are mostly perennials growing in grasslands, rough ground, waysides, hedgerows and woods.

Interpretative value

Grains may have been smothered by midden deposits, present in waste fodder and bedding, dung or cereal processing waste, or amongst grassy vegetation used as toilet wipes, floor covering, etc.

Family: POACEAE
Latin name: *Hordeum vulgare sens. lat.*
Common name: barley
Anatomical element: grain

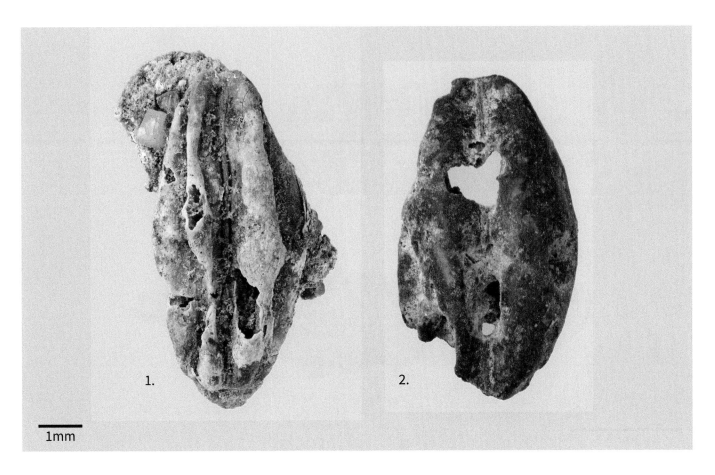

1mm

Image description

1. Example 1; grain, ventral view, palea and lemma not preserved (concretion across dorsal surface) [B125].

2. Example 2; grain, ventral view, palea and lemma not preserved (dorsal surface not preserved) [B129].

Key diagnostic features and separation from similar taxa

- Dorsally-compressed spindle-shaped grain with shallow furrow closely comparable with modern reference material. Husks (palea and lemma) not usually preserved.

- Because of distinctive dorsally compressed tapered shape other cereal grains unlikely to be mistaken for barley if sufficiently well-preserved.

- Possibility of distortion and partial preservation in large dry-seeded items such as cereal grains fairly high. Identification beyond *Hordeum* sp. not likely to be possible.

Examples of archaeological sites

Late Saxon cesspits at Discovery Centre, Winchester [HA1]; medieval cesspit at 70, Stour Street, Canterbury [KT1].

Modern ecology

Cultivated crop plant, grown in British Isles from Neolithic onwards. Tolerates wide variety of soils, including poorer, dry soils and copes with coastal conditions better than other cereals (Zohary *et al* 2013). Range of uses includes fodder, bread flour, whole grains in soups and stews and main cereal used to provide malt for brewing since medieval times.

Interpretative value

Whole grains may have been used to decorate bread, contained and consumed within bread if poorly milled, or cooked in soups and pottages. On middens could have derived from waste fodder and animal bedding or faecal waste. *Triticum* sp. and *Avena* sp. caryopses have also occasionally been preserved by mineralisation, for example in Roman cesspits at Silchester, Hants (Robinson 2006).

Order: Diptera
Family: PSYCHODIDAE
Latin name: *Psychoda alternata* Say.
Common name: Drain fly or trickling filter fly
Anatomical element: Puparium

1.

2.

1mm

Image description

1. Whole puparium, ventral view [E1].
2. Whole puparium, lateral view [E1].

Key diagnostic features and separation from similar taxa

- Distinct and separate head and thoracic capsule often with impressions forming legs and wings.
- Tubular and segmented hind section (abdomen).
- Abdominal segments which have a series of thick seti (or bristles) thickened to form 'spikes' on puparium along the sides and across both ventral and dorsal surfaces.
- Can be confused with a range of similar flies, particularly *Scatopse notata* (L.), if the arrangement of the 'spikes' on the abdomen is not clear.

Examples of archaeological sites

Mineralised from pits from Medieval Southampton French Quarter (Smith, D. 2009) and Medieval Finzel's Reach, Bristol (Smith, D. 2017).

Modern ecology

The 'drain fly'. This fly species is normally associated with bacterial mats and sludge in wet or damp areas and decaying vegetation (Smith, K.G.V. 1989). Today, it is commonly associated with clogged plumbing, pooled water and sewage.

Interpretative value

Normally associated with cesspits in the archaeological record (Smith, D. 2013).

Order: Diptera
Family: SYRPHIDAE
Latin name: *Eristalis tenax* (L.)
Common name: Rat-tailed maggot or drone fly
Anatomical element: Whole puparium

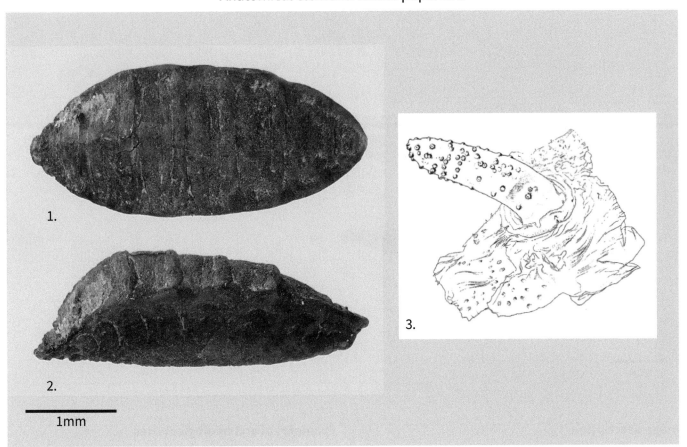

1mm

Image description

1. Whole puparium, dorsal view [E2].
2. Whole puparium, lateral view [E2].
3. Image taken from Phipps (1988) and drawn by Alan Robertson, showing example of anterior spiracles.

Key diagnostic features and separation from similar taxa

- Puparia relatively small 'flat-bottomed' and arched at posterior end.
- Puparia clearly segmented and with 'pie crust' margin and ridged side.
- Anterior spiracle horns are very characteristic (resemble small bits of coral around 1.5 mm long, normally amber in colour) and are often found separated from the puparia.

Examples of archaeological sites

Common in some archaeological sites. Pits from Medieval Southampton, French Quarter (Smith, D. 2009). Two stone-lined latrines, Medieval Free School Lane, Leicester (Smith, D. 2008) and pits from Medieval Finzel's Reach, Bristol (Smith, D. 2017).

Modern ecology

E. tenax is a specialist in pools of stagnant water with a high organic content. It tolerates pollution and low levels of oxygenation and is common in sewage lagoons and cesspools (Smith, K.G.V. 1973).

Interpretative value

E. tenax is a classic indicator of liquid cess and pools of faecal material in the archaeological record (Skidmore 1999; Smith, D. 2013).

Order: Diptera
Family: SEPSIDAE
Latin name: *Sepsis* **spp.**
Common name: Dung fly
Anatomical element: Puparium

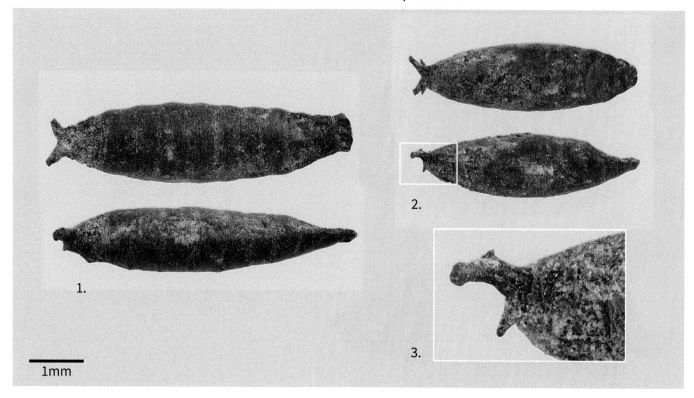

1mm

Image description

1. Whole puparium, dorsal and lateral view, specimen 1 [E3].
2. Whole puparium, dorsal and lateral view, specimen 2 [E3].
3. Enlargement of posterior end of puparium [E3]

Key diagnostic features and separation from similar taxa

- Relatively large slightly flattened puparia often with emarginate edges.
- Anal spiracles mounted on a pair of long tube usually with two associated 'spikes' giving the appearance of a 'crown of thorns'.
- Anterior end is quite flattened giving a wedge shape to the first segment bearing the mouth parts.
- Segments of the body often bear very rough ridged surfaces and/ or large numbers of small spikes or a roughened surface.
- Relatively easy to identify to genus level, but very difficult to identify to species level.

Examples of archaeological sites

Very common on a wide range of Roman and Medieval sites. Pits from Medieval Southampton, French Quarter (Smith, D. 2009). Two stone-lined latrines, Medieval Free School Lane, Leicester (Smith, D. 2008). Roman and Medieval deposits Causeway Lane, Leicester (Skidmore 1999) and Medieval Finzel's Reach, Bristol (Smith, D. 2017).

Modern ecology

Sepsis flies occur in a wide range of liquid animal dung and cess (Smith, K.G.V. 1989).

Interpretative value

Normally associated with cesspits in the archaeological record (Smith, D. 2013), but can be recovered in some numbers from a range of other archaeological deposits (Smith, D. 2012). Several of the *Sepsis* species appear to utilize the dung of specific herbivores or are limited to human cess (Smith, K.G.V. 1989). Further work on the identification of species of this genus in the archaeological record is warranted.

Order: Diptera
Family: SPHAEROCERIDAE
Latin name: *Thoracochaeta zosterae* (Hal.)
Common name: Seaweed or cesspit fly
Anatomical element: Whole puparium

1mm

Image description

1. Whole puparium, dorsal and lateral view [E4].
2. Pupa from inside a puparium, ventral and lateral view 'light form' [E4].
3. Enlargement of posterior end [E4].
4. Enlargement of side of puparium [E4].

Key diagnostic features and separation from similar taxa

- Long tubular flask shaped puparium.
- Heavily segmented larvae with distinctive 'pie crust' margins on the sides (Figure 4).
- Anal spiracles mounted on a bifurcated tube on the posterior end (Figure 3).
- Lighter coloured forms often show the 'shadow' of the developing legs of the pupa within the puparium (Figures 2).
- Can be identified with a large degree of confidence.

Examples of archaeological sites

Widely found in urban faecal deposits, particularly Roman and medieval cesspits, both as waterlogged and mineralised remains. Mineralised in pits from Medieval Southampton, French Quarter (Smith, D. 2009). Two stone lined latrines, Medieval Free School Lane, Leicester (Smith, D. 2008). Roman and Medieval deposits Causeway Lane, Leicester (Skidmore 1999) and Medieval Finzel's Reach, Bristol (Smith, D. 2017).

Modern ecology

Species is today mainly found with seaweed on the coast where the larvae are associated with wet weed and the puparia with dry rack (Webb *et al* 1998; Belshaw 1989).

Interpretative value

Archaeologically, there is a clear link between this species and cess and rubbish pits (Belshaw 1989; Skidmore 1999; Smith, D. 2013) and it is suggested that *T. zosterae* is essentially an indicator species for this type of deposit. Often *T. zosterae* occurs as several hundred individuals from a single 5-10 litre sample. It is thought they occur in semi-fluid 'filth', which contains a high level of dissolved salts and organic matter derived from urine and cess It is suggested that similar conditions either in terms of moisture, chemistry or repetitive wetting and drying occur in stands of seaweed by the coast (Belshaw 1989; Smith, D. 2013) also occur in these pits.

Order: Diptera
Family: CALLIPHORIDAE
Latin name: *Calliphora vomitoria* (L.)
Common name: Rural bluebottle
Anatomical element: Puparium

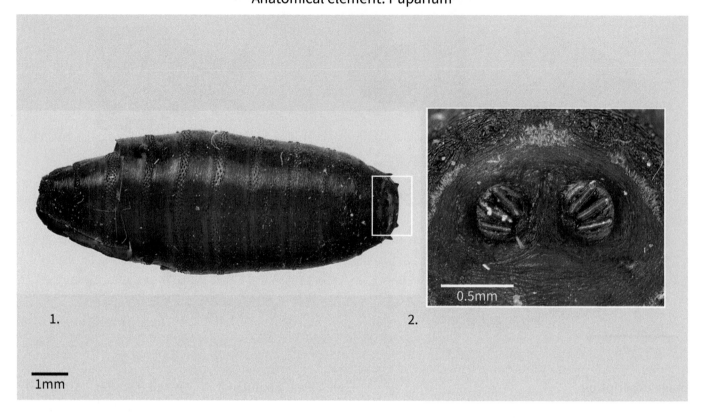

1.

2.

0.5mm

1mm

Image description

1. Whole puparium (modern specimen), ventral view [E5].
2. Whole puparium (modern specimen), posterior view – showing posterior spiracles [E5].

Key diagnostic features and separation from similar taxa

- Large cylindrical puparia with a rough, but shiny surface between the segments.

- Cuticle surface of the segments between the 'belts' finely ridged. The 'belts' between the segments often have raised ambulatory welts consisting of chains of raised spikes.

- Posterior segment has a 'crown of thorns' of spikes around the posterior spiracles (see right side of puparium – Figure 1).

- Posterior spiracles very characteristic. Raised on shallow platforms. Usually round in shape. A small button is adjacent to three long straight slits.

- Identification is very easy provided the posterior spiracles are preserved and clear.

Examples of archaeological sites

Often occur in limited numbers in a wide range of waterlogged contexts from Roman and Medieval settlement (Smith, D. 2012, 2013). Mineralised examples have been recovered from Medieval Finzel's Reach, Bristol (Smith, D. 2017).

Modern ecology

The rural bluebottle is a carrion feeder, usually associated with cadavers in the early stages of decay or with decaying meat and/ or food waste (Smith, K.G.V. 1973, 1989).

Interpretative value

Thought to be associated with decaying corpses and kitchen waste in the archaeological record.

Order: Diptera
Family: FANNIIDAE
Latin name: *Fannia ?scalaris* **(Fab.)**
Common name: Latrine fly
Anatomical element: Puparium

1mm

Image description

1. Whole puparium, ventral view [E6].
2. Whole puparium, lateral view [E6].
3. Enlargement of distal end showing posterior spiracles and 'crown' [E6].

Key diagnostic features and separation from similar taxa

- Distinctive 'segmented' body with the remains, often 'stumps' of filiform processes on sides.

- Distinctive 'crown' of 'stumps' or processes at distal end with two palmate spiracles in the central area.

- Can be easily confused with the 'lesser housefly' *F. canicularis* in which the filiform processes are narrower and less 'feathered'. The processes are more commonly preserved and identifiable in waterlogged material. Often it is assumed that specimens from cesspits are *F. scalaris* but identification to the level of *F. ?scalaris* may be safer.

Examples of archaeological sites

Mineralised from pits at Medieval Southampton French Quarter (Smith, D. 2009). Two stone lined latrines, Medieval Free School Lane, Leicester (Smith, D. 2008), Roman and Medieval deposits Causeway Lane, Leicester (Skidmore 1999) and Medieval Finzel's Reach, Bristol (Smith, D. 2017).

Modern ecology

Associated primarily with semi-liquid faeces and urine often of humans or pigs (Smith, K.G.V. 1989).

Interpretative value

A strong indicator for the presence cess, often human. Has been suggested that it is often present in cesspits which are fluid or have standing water since the filiform processes are thought to allow flotation (Smith, K.G.V. 1989).

Order: Diptera
Family: MUSCIDAE
Latin name: *Musca domestica* L.
Common name: The house fly
Anatomical element: Puparium

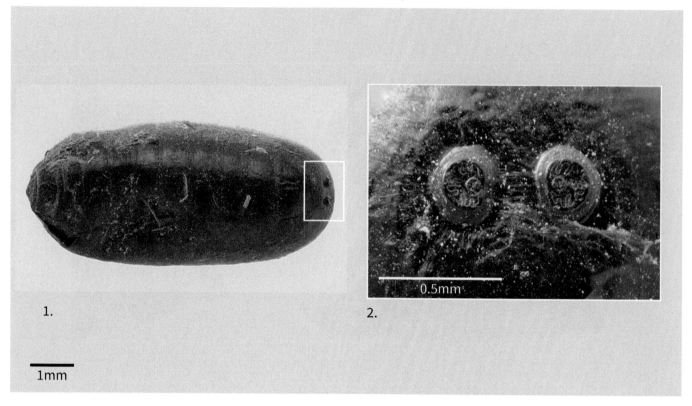

1.

2.

0.5mm

1mm

Image description

1. Whole puparium, dorsal view [E7].
2. Whole puparium, posterior view – showing posterior spiracles [E7].

Key diagnostic features and separation from similar taxa

- Large cylindrical puparia usually with a smooth and shining surface.
- Cuticle surface of the segments between the 'belts' is smooth. 'Belts' between the segments consisting of long thin ridges without teeth or welts on ventral surface.
- Posterior spiracles very characteristic. Raised on short platforms. Flattened oval in shape (shaped like an orange segment). 'Button' feature on spiracles is offset towards the inner margin and surrounded by two or three sinuous 'serpentine' slits.
- Identification is very easy provided the posterior spiracles are preserved and clear.

Examples of archaeological sites

Very common in a wide range of waterlogged contexts from human settlement in the archaeological record, including prehistoric periods (Panagiotakopulu and Buckland 2018). Mineralised examples have been recovered Medieval Finzel's Reach, Bristol (Smith, D. 2017).

Modern ecology

The 'house fly'. This is one of the most common flies in human housing and other settlement deposits in the archaeological record. It is a true synanthropic species associated with human food, rubbish and waste in towns (Smith, K.G.V. 1973, 1989). In the countryside, it often is associated with stable waste and animal dung (Smith, K.G.V. 1973, 1989).

Interpretative value

It is a very strong indicator for human settlement, waste deposits and stabling materials in the archaeological record (Smith, D. 2013; Panagiotakopulu and Buckland 2018).

Order: Diptera
Family: MUSCIDAE
Latin name: *Muscina stabulans* (Fall.)
Common name: The false stable fly
Anatomical element: Puparium

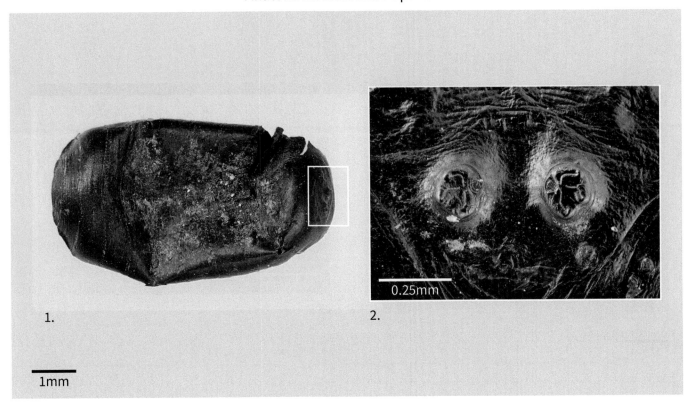

1.

2.

1mm

0.25mm

Image description

1. Whole puparium, dorsal view [E8].

2. Whole puparium, posterior view – showing posterior spiracles [E8].

Key diagnostic features and separation from similar taxa

- Large cylindrical puparia with a mat surface between the segments.

- Cuticle surface of the segments between the 'belts' finely ridged, resulting in a mat surface. 'Belts' between the segments have raised ambulatory welts giving a shallow saw-toothed margin (see left side of puparium in Figure 1).

- Posterior spiracles very characteristic. Raised on shallow platforms. Round or even roughly pentagonal in shape. 'Button' offset towards the inner margin and surrounded by three curved slits which are arranged roughly into 'a head of a trident' shape.

- Identification is very easy provided the posterior spiracles are preserved and clear.

Examples of archaeological sites

Common in a wide range of waterlogged contexts from Roman and Medieval settlement (Smith, D. 2012, 2013). Mineralised examples have been recovered Medieval Finzel's Reach, Bristol (Smith, D. 2017).

Modern ecology

The 'false stable fly' is common in housing, but is also associated with stables, byres and decaying organic matter (Smith, K.G.V. 1973, 1989).

Interpretative value

It is a very strong indicator for human settlement, waste deposits and stabling materials in the archaeological record.

Order: Diptera
Family: MUSCIDAE
Latin name: *Stomoxys calcitrans* (L.)
Common name: Stable fly or biting house fly
Anatomical element: Puparium

1.

1mm

2.

0.25mm

Image description

1. Whole puparium, dorsal view [E9].

2. Whole puparium, posterior view – showing posterior spiracles [E9].

Key diagnostic features and separation from similar taxa

- Large cylindrical puparia with a shiny surface between the segments

- Cuticle surface of the segments between the 'belts' slightly ridged or wrinkled particularly towards the side, but still often with a shining surface. The 'belts' between the segments often have raised ambulatory welts, producing a shallow saw-toothed margin.

- Posterior segment on dorsal surface usually has three raised lines of bumps running from the margin of the segment towards the posterior end (see right side of puparium in Figure 1).

- Posterior spiracles very characteristic. Raised on shallow platforms. Usually a rounded triangle in shape. 'Button' placed in the center of the platform and surrounded by three curved (sometimes serpentine) slits.

- Identification is very easy provided the posterior spiracles are preserved and clear.

Examples of archaeological sites

Common in a wide range of waterlogged contexts from Roman and Medieval settlement (Smith, D. 2012, 2013). Mineralised examples have been recovered from Medieval Finzel's Reach, Bristol (Smith, D. 2017).

Modern ecology

The 'stable fly' is common in housing, but is mainly associated with stables, byres and decaying organic matter (Smith, K.G.V. 1973, 1989). The adults feed on the blood of a range of stock animals, particularly cattle and horses. It is one of the few flies that actively bites humans and is most likely responsible for the hard welts left on exposed skin after walks in the country.

Interpretative value

It is a very strong indicator of human settlement and can suggest the presence of stabling material (Kenward and Hall 1997).

Order: Diptera
Family: HIPPOBOSCIDAE
Latin name: *Melophagus ovinus* (L.)
Common name: Sheep Ked
Anatomical element: Puparium

1.

1mm

Image description

1. Whole puparium, dorsal view [E10].

Key diagnostic features and separation from similar taxa

- Very distinctive large globular puparium with smooth surface
- Two rows of bowl-shaped depressions on dorsal surface, with six on each side.
- The anterior end often bears a large 'circular scar' with a raised point in the middle.

Examples of archaeological sites

Rare in mineralised material but has been recovered from Finzel's Reach, Bristol (Smith, D. 2017). Has been recovered widely from a number of waterlogged urban sites.

Modern ecology

The adult sheep ked is a well-known external parasite of sheep, which constantly bites and draws blood from the skin. The female retains the larvae within its own body until it is almost ready to pupate. The pupa is attached to the wool by a sticky thread.

Interpretative value

Has been widely used in the archaeological record to suggest the presence of sheep and, more importantly, wool processing, cloth working and dyeing (eg Buckland and Perry 1989; Hall and Kenward 2003).

Order: Coleoptera
Family: CUCUJIDAE
Latin name: *Oryzaephilus surinamensis* (L.)
Common name: Saw-toothed grain beetle
Anatomical element: whole and fragmented adults

1mm

Image description

1. Whole adult beetle (modern specimen), dorsal view.

2. Individual head, thoraxes and elytra (charred specimens) [E11].

3. Individual thorax and elytra (waterlogged specimen) [E11]

Key diagnostic features and separation from similar taxa

- Heavily punctured triangular head with sharp hind angles.

- Distinctive thorax with 6 'teeth' on each side.

- Elytra long and narrow with raised ridges and prominent punctures.

Examples of archaeological sites

Common on Roman and Medieval waterlogged archaeological sites (Smith, D. 2012). Can occur in limited numbers in cesspits and as mineralised material (Smith, D. 2013). Recovered as charred individuals from Roman ditches at Northfleet, Kent (Smith, D. 2017). Mineralised in pits from Medieval Southampton, French Quarter (Smith, D. 2009).

Modern ecology

The 'saw-toothed grain beetle' is a common pest of stored products, particularly decayed grain. It is normally associated with grain stores, mills and bakeries.

Interpretative value

The saw-toothed grain beetle has been encountered in large numbers in a range of urban and rural features in the archaeological record from the early Roman period onwards. It has been used to indicate the presence of considerable quantities of stored grain at a number of locations (eg Smith and Kenward 2011, 2013). It is also thought to find its way into cesspits as the result of having been consumed in foodstuffs (Smith 2013). It is probably through this latter route that it enters the deposit in which it becomes mineralised.

Order: Coleoptera
Family: CURCULIONIDAE
Latin name: *Sitophilus granarius* (L.)
Common name: Granary weevil
Anatomical element: whole and fragmented adults

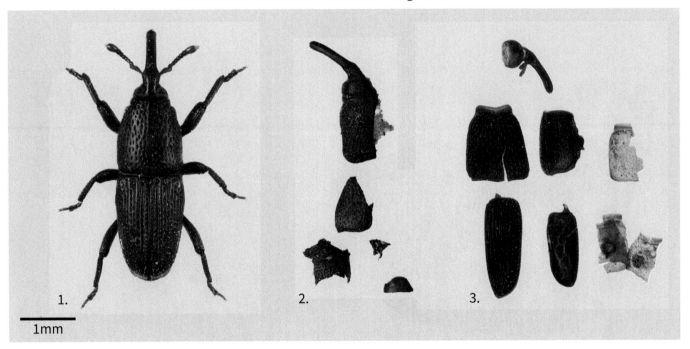

1mm

Image description

1. Whole puparium (modern specimen), dorsal view.
2. Individual head, thoraxes and elytra (charred specimen) [E12].
3. Individual thorax and elytra (waterlogged specimen) [E12].

Key diagnostic features and separation from similar taxa

- Head with long curved tubular rostrum or 'snout'. This is widest just before the eyes, where the antennae are attached. This area has a distinctive set of three rows of punctures.

- Stout cone shaped thorax with a strong ridge of material around the neck. The thorax is covered in large rounded punctures.

- Elytra relatively short with blunted posterior (apical) ends and a strong tooth on the anterior end, near to the scutellum insertion (this is the triangular or slot shape 'cut out' where the 'hinge' between the two elytron would sit). The elytra are extremely ridged with lines of strong punctures between the ridges.

Examples of archaeological sites

Common on Roman and Medieval waterlogged archaeological sites particularly where there are stores or warehouses (Smith, D. 2012). Can occur in limited numbers in cesspits and as mineralised material (Smith, D. 2013). Recovered as charred individuals from Roman ditches at Northfleet, Kent (Smith, D. 2017). Mineralised in pits from Medieval Southampton French Quarter (Smith, D. 2009).

Modern ecology

The granary weevil is a common primary pest of stored grain where it attacks whole grain which is only slightly damp. It is therefore normally associated with grain stores, mills and bakeries.

Interpretative value

The granary weevil is often encountered in large numbers in a range of urban and rural features in the archaeological record from the early Roman period onwards. It has been used to indicate the presence of quantities of stored grain at a number of locations and in a wide range of deposits (eg Smith and Kenward 2011, 2013). It probably finds its way into cesspits as the result of having been consumed in foodstuffs (Smith 2013). It is probably through this latter route that it enters the deposits in which it becomes mineralised.

BRAN CURLS EMBEDDED IN FAECAL CONCRETIONS

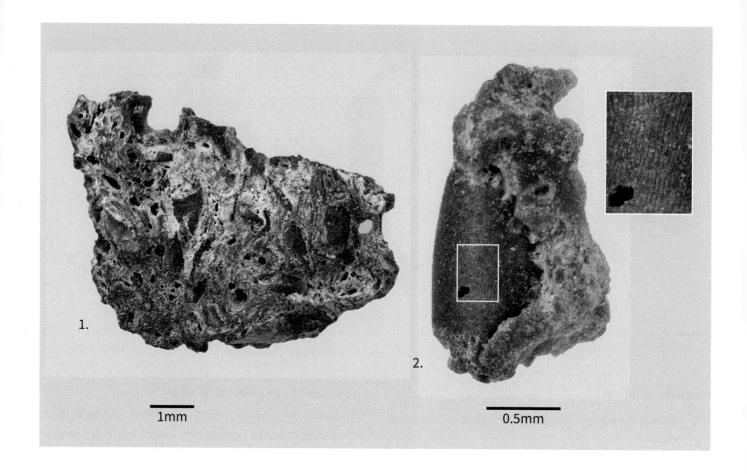

1mm

0.5mm

Image description

1. Faecal concretion containing frequent bran curls [B147].

2. High magnification image of single bran curl showing characteristic elongated cells of the pericarp [B165].

Key diagnostic features and separation from similar taxa

- Presence of small curled fragments of cereal bran within amorphous amber coloured matrix is indicative of faecal concretions.

- Individual bran curls sometimes have faint but distinctive cross-cell patterns on surface where thin vertically and horizontally elongated cell layers superimposed.

- Fragments of pulse seed coat and fruit skin tend to be larger and thicker with their own distinctive characteristics (see pages 75 and 76).

Examples of archaeological sites

Widely found in faecal deposits but not easy to quantify. Rough estimates of percentage of faecal concretions in a sample can be made by visually scanning a number of petri-dishes of residue.

Interpretative value

Principal indicators for the presence of human faecal waste (Smith 2013). Appearance of mineralised concretions varies considerably depending on preservation conditions (sometimes knobbly or angular). Not always well-enough preserved to confirm that bran is present.

STRAW/LARGE GRASS CULM AND INFLORESCENCE FRAGMENTS

5mm

1mm

Image description

1. Poaceae culm fragment, terete (rounded) and hollow, plus high magnification image showing elongated cells [B126].

2. Concretion containing matted straw/grass fragments [B149].

3. Indeterminate Poaceae spikelet [B128].

4. Poaceae culm node [B130].

5. Second example of culm node from above showing characteristic channels that once accommodated vascular bundles [B131].

Key diagnostic features and separation from similar taxa

- Stems of Poaceae are usually terete (sometimes flattened), often hollow with long narrow epidermal cells.

- Juncaceae (see *Juncus* sp. page 62) usually have terete, ridged, solid stems filled with pith, no nodes and narrow elongated epidermal cells.

- Cyperaceae often triangular in cross-section with pith, no nodes and narrow elongated epidermal cells.

- Unlikely to be able to distinguish between cereal straw and large grass stems.

Examples of archaeological sites

Late Bronze Age midden at Potterne [WT1]; Late Bronze Age midden at East Chisenbury [WT3].

Interpretative value

Fragments of mineralised straw/grass often main indicators of faecal deposits in pits, though animal bedding/fodder could also produce this type of assemblage. Probably used to soak up liquids, reduce odours, also as toilet wipes and floor covering. If culm fragments numerous and matted straw/grass present, may be worthwhile roughly estimating as a percentage of assemblage.

RUSH (*JUNCUS* SP.) CULM FRAGMENTS

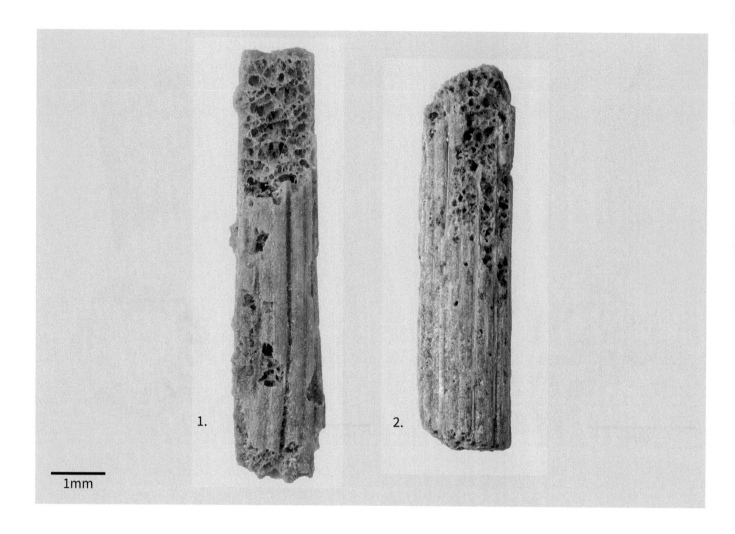

1mm

Image description

1. Example 1; fragment of stem, visible pith continuous [B24].
2. Example 2; fragment of stem, visible pith continuous [B175].

Key diagnostic features and separation from similar taxa

- Deeply ridged terete stems with central pith (more deeply furrowed when dehydrated).
- Where pith continuous cannot be identified to species level, but where clearly interrupted likely to be *J. inflexus.*
- Poaceae stems (straw and grasses) generally smoothly cylindrical and hollow (see Page 61). Sedge stems usually triangular in section with pith.

Examples of archaeological sites

Late Bronze Age midden at East Chisenbury [WT3]; frequent in late Saxon cesspit at Northgate House, Winchester [HA1].

Modern ecology

Juncus spp. primarily grow in damp to wet habitats such as marshes, bogs, fens, heaths, flushes, damp grasslands, dune slacks, saltmarshes.

Interpretative value

Often recovered from cesspits in moderate numbers amongst more abundant straw/grass stem fragments. Likely to have been used as toilet wipes, floor covering or to reduce odours and soak up liquids.

PEA/BEAN (*PISUM SATIVUM/VICIA FABA*) SEED COAT FRAGMENTS

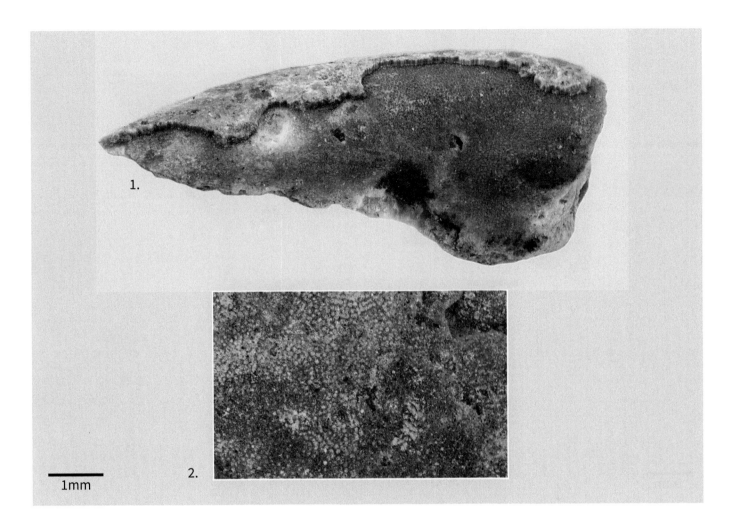

1mm

Image description

1. Fragment of pulse seed coat showing columnar cells of palisade layer over hourglass-cells in layer below [B162].

2. High magnification detail of cells in outer epidermis, showing polygonal tops of palisade cells [B161].

Key diagnostic features and separation from similar remains

- Thickness of seed coat and distinctive epidermal cell pattern enable fragments to be picked out from samples.

- Often preserved as large curls of pulse seed coat, as distinct from folded fragments of thinner fruit skin (see Page 76).

- Can only be identified further where hila are also present.

Examples of archaeological sites

Middle Saxon cesspits, The Deanery, Southampton [HA4]; late Saxon and medieval cesspits at Discovery Centre, Winchester [HA1].

Interpretative value

Useful indicator of cesspits and faecal deposits (Smith 2013). Important to roughly quantify fragments to determine how important pulses were in diet.

CF. FRUIT EXOCARP (SKIN)

1mm

1.

2.

Image description

1. Example 1; exocarp within faecal material. Probably *Prunus* sp. in this example as sloe and plum stones were abundant [B42].

2. Example 2; high magnification image of exocarp of whole *Prunus spinosa* fruit (confirmed identification; see *Prunus spinosa*, page 9) [B40].

Key diagnostic features and separation from similar remains

- Faintly dimpled polygonal cells visible in some areas beneath waxy cuticle.

- Difficult to identify isolated fragents as fruit exocarp with certainty due to lack of diagnostic features.

- Cell patterns less distinctive than pulse seed coat fragments. Fruit exocarp tends to be thinner and more often folded, large fragments of thicker pulse seed coat often rolled.

Examples of archaeological sites

Roman cesspit at 49, St Peter's Street, Canterbury [KT4]; mid Saxon pits at The Deanery, Southampton [HA4].

Interpretative value

Limited due to difficulties in identification, though recovery of possible fruit skin does suggest presence of either faecal material or domestic waste.

DICOTYLEDONOUS PLANT ROOTS

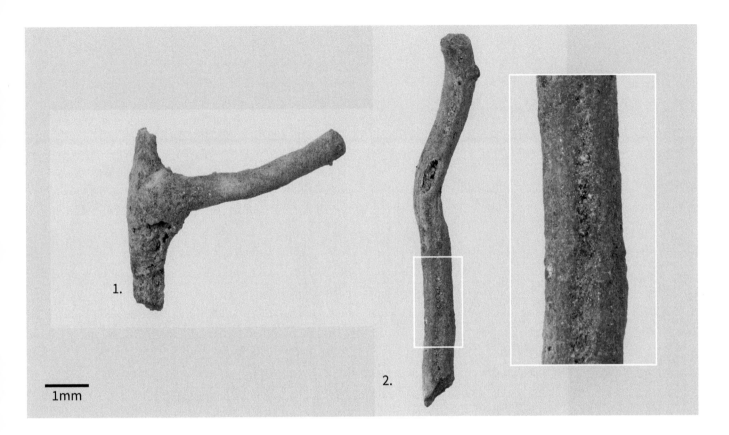

1mm

Image description

1. Branched root [B25].
2. Unbranched root with high magnification image of cell pattern [B176].

Both roots come from midden at Potterne [WT1]. Over 300 roots (c. 10% sub-sample) identified by T. Lawrence (Jodrell Laboratory, Kew Gardens) as belonging to dicotyledonous plants, monocotyledons not represented (Carruthers 2000). See McCobb *et al* (2003) for scanning electron micrographs of vascular vessels (Fig. 3c and 3d), discussion of findings and most likely explanation of process of fossilisation accounting for preservation of only dicotyledonous roots at Potterne (page 1279).

Key diagnostic features and separation from similar remains

- Thin, twisted, occasionally branched roots with indistinct cell pattern.
- In roots examined by McCobb *et al* (ibid) only the central vascular cylinder (stele) had become preserved by mineralisation (only likely to be determined using SEM).
- Occasional roots might be difficult to recognise with certainty but where frequent the size and twisted forms are distinctive.

Examples of archaeological sites

Abundant in lower layers of Late Bronze Age midden and pre-midden sediments at Potterne [WT1]; occasional in Iron Age ditch and pits at Battlesbury Bowl [WT2].

Interpretative value

Distribution at Potterne (Carruthers 2000) helped to confirm that mineralisation had occurred in situ. Large numbers of fragments could indicate periods of stabilisation of vegetation cover, followed by smothering with midden material. Roots unlikely to be preserved in cesspits unless uprooted plants discarded, or vegetation cover rapidly established, rooting into top of pits. Further SEM analysis of roots may be worthwhile to determine whether monocotyledonous roots are sometimes preserved under different conditions.

EARTHWORM COCOON

1mm

Image description

1. Earthworm cocoon with polar tuft [B26].
2. Juvenile worm in cocoon [B108].

Key diagnostic features and separation from similar remains

- Smooth surface with no clear cell pattern.
- Polar tuft not always preserved but usually either depression or small nipple at either end of sub-spherical cocoon.
- Size varies, possibly several species represented, but cannot be identified further than 'probably… native British lumbricids' (Piearce *et al* 1992).
- Distinguished from spherical *Brassica/Sinapis* sp. seeds (see page 20) by absence of clear cell pattern and polar tufts.

Examples of archaeological sites

Late Bronze Age features below midden, Potterne [WT1]; mid Saxon cesspit, The Deanery, Southampton [HA4].

Modern ecology

Earthworm species suggested by Piearce (ibid) for specimen from Potterne occupy wide range of habitats with exception of very acidic soils or very wet places such as mires.

Interpretative value

Earthworms often abundant where deposits are rich in decaying organic matter, such as in middens and compost heaps. Preservation of large numbers of complete cocoons suggests large scale deposition took place producing conditions that smothered cocoons and prevented hatching.

RODENT DROPPINGS

1mm

Image description

1. Large dropping with two pinched ends, most likely rat [B158].
2. Small dropping with pinched end, probably mouse [B54].

Key diagnostic features and separation from similar remains

- Elongated oblong with at least one pointed end, often pinched and slightly twisted.
- Surface not uniform and no cell structure visible.
- Often pale colouration.
- Size and form varies depending on rodent. Online pest control websites provide the following aids to identification; mouse droppings - small (3-6mm) pinched at one or both ends; black rat - large (7-13 mm), spindle-shaped with two pointed ends; brown rat - often larger (7-19mm), cylindrical with more rounded ends. However, accurate identification may not be possible on basis of individual droppings and other rodents cannot be ruled out.

Examples of archaeological sites

Mid Saxon cesspits at St Mary's Stadium, Southampton [HA5]; Saxon and medieval cesspits at Northgate House and Discovery Centre, Winchester [HA1].

Modern ecology

Typically found wherever food is stored and prepared, particularly in urban locations where large populations of rodents can become established. Also close to waterways, around farm buildings and where animals are housed.

Interpretative value

Large numbers of droppings provide some indications of unsanitary conditions present on a site. Cesspits containing large numbers of droppings may have been abandoned and colonised by rodents, or material swept up from floors may have been deposited in pits.

UNIDENTIFIED MINERALISED 'NODULES'

1.

2.

1mm

Image description

1. Nodules showing size range and variation in form, plus high magnification image showing surface detail [B27].

2. Broken nodule showing typical internal morphology [B79].

Key diagnostic features and separation from similar remains

- Sub-spherical and usually dimpled on one side.

- Rough surface but no obvious cell structure to indicate that they are of biological origin.

- Wide variation in size, can be abundant in cesspits and middens.

- Usually hollow when broken open with a texture suggesting some type of crystallisation has taken place.

Examples of archaeological sites

Widely found where phosphatic mineralisation has taken place, eg Late Bronze Age midden at Potterne [WT1]; Saxon and medieval cesspits at Discovery Centre and Northgate House, Winchester [HA1].

Interpretative value

Indicators that mineralisation has taken place. Since some are very large they are often seen during soil processing (often mistaken for peas until examined under the microscope) so can alert archaeologists to presence of mineralisation. Although formation and significance not yet understood (Carruthers 1989; Matt Canti, Historic England, pers. comm.) are worth recording in case future advances find them to be useful.

Appendix I: Voucher Specimens

Voucher No.	Latin name	Common name & anatomical element	Sample no.	Context no.	Deposit type	Period	Site code	Site	Unit	Page in Guide
B1	*Vicia faba* L.	field bean hilum	109	1026	cesspit	late Saxon	HA1	Discovery Centre, Winchester, Hants	Oxford Archaeology	6
B2	*Pisum sativum* L.	pea hilum	9	1228	cesspit	Saxo-Norman	BD1	Willington to Steppingley pipeline	Network Archaeology	7
B4	*Montia fontana* L.	blinks seed	567	2606	midden	Late Bronze Age	WT1	Potterne, Wilts	Wessex Arch	30
B5	*Conium maculatum* L.	hemlock seed	26	509	cesspit	C13th-C16th	LD2	Aldersgate, London City Ditch	Pre-Construct Archaeology	52
B6	*Aphanes arvensis* L.	parsley piert seed	2013	4192	ditch 4043	Iron Age	WT2	Battlesbury Bowl, Wilts	Wessex Arch	13
B7	*Hyoscyamus niger* L.	henbane seed	567	2606	midden	Late Bronze Age	WT1	Potterne, Wilts	Wessex Arch	36
B8	*Urtica dioica* L.	common nettle seed	2013	4192	ditch 4043	Iron Age	WT2	Battlesbury Bowl, Wilts	Wessex Arch	15
B9	*Thlaspi arvense* L.	field penny-cress seed	555	2209	midden	Late Bronze Age	WT1	Potterne, Wilts	Wessex Arch	22
B10	*Urtica urens* L.	small nettle seed	2013	4192	ditch 4043	Iron Age	WT2	Battlesbury Bowl, Wilts	Wessex Arch	16
B11	*Agrostemma githago* L.	corn cockle seed	257	2342		Anglo-Norman	HA1	Discovery centre, Winchester, Hants	Oxford Archaeology	27
B12	*Fallopia convolvulus* (L.) Á. Löve	black bindweed fruit	301	3045	earth floors	late medieval	KT1	Stour Street, Canterbury	Canterbury Archaeological Trust	23
B13	*Silene* sp.	campion seed	301	3045	earth floors	late medieval	KT1	Stour Street, Canterbury	Canterbury Archaeological Trust	28
B14	*Daucus carota* L.	carrot fruit/seed	301	3045	earth floors	late medieval	KT1	Stour Street, Canterbury	Canterbury Archaeological Trust	55
B15	*Lithospermum arvense* L.	field gromwell seed	2193	5732	pit 5592	Iron Age	WT2	Battlesbury Bowl, Wilts	Wessex Arch	34
B16	*Linum catharticum* L.	fairy flax seed	2193	5732	pit 5592	Iron Age	WT2	Battlesbury Bowl, Wilts	Wessex Arch	18
B17	*Rumex acetosella* L.	sheep's sorrel seed	2193	5732	pit 5592	Iron Age	WT2	Battlesbury Bowl, Wilts	Wessex Arch	24
B18	*Rumex crispus*-type	dock seed	2193	5732	pit 5592	Iron Age	WT2	Battlesbury Bowl, Wilts	Wessex Arch	25
B19	*Anagallis arvensis*-type	scarlet pimpernel-type	2193	5732	pit 5592	Iron Age	WT2	Battlesbury Bowl, Wilts	Wessex Arch	31
B20	*Vitis vinifera* L.	grape pip			garderobe 4051	early 13th/14th C	BK1	Jennings Yard, Windsor, Berks	Wessex Arch	5
B21	*Lapsana communis* L.	nipplewort fruit & seed			base of City Ditch	13th-16th C	LD2	London City Ditch, Aldersgate	Pre-Construct Archaeology	44
B22	*Stellaria* sp.	stitchwort seed	2144	4603	pit 4598	Iron Age	WT2	Battlesbury Bowl, Wilts	Wessex Arch	26
B23	*Myosotis* sp.	forget-me-not seed	2165	4742	pit 4641	Iron Age	WT2	Battlesbury Bowl, Wilts	Wessex Arch	35
B24	*Juncus* sp.	rush stem	180	2377	pit 2373	late Saxon	HA1	Northgate House, Winchester, Hants	Oxford Archaeology	74
B25		dicotyledonous root	4	50	midden	Late Bronze Age	WT1	Potterne, Wilts	Wessex Arch	77
B26		worm cocoon	2162	5162	pit 5149	Iron Age	WT2	Battlesbury Bowl, Wilts	Wessex Arch	78
B27		nodules	2062	4460	pit 4458	Iron Age	WT2	Battlesbury Bowl, Wilts	Wessex Arch	80
B28	*Papaver* sp.	poppy seed	2013	4192	ditch 4043	Iron Age	WT2	Battlesbury Bowl, Wilts	Wessex Arch	2
B30	*Linum catharticum* L.	fairy flax seed	2013	4192	ditch 4043	Iron Age	WT2	Battlesbury Bowl, Wilts	Wessex Arch	18
B31	*Lithospermum arvense* L.	field gromwell seed	2125	4674	pit 4667	Iron Age	WT2	Battlesbury Bowl, Wilts	Wessex Arch	34
B32	*Ranunculus acris/bulbosus/repens*	buttercup seed	2120	4600	pit 4598	Iron Age	WT2	Battlesbury Bowl, Wilts	Wessex Arch	4
B33	*Torilis* sp.	hedge-parsley fruit / seed			garderobe 4051	early 13th/14th C	BK1	Jennings Yard, Windsor, Berks	Wessex Arch	54
B35	*Ficus carica* L.	fig fruit	802	841	cesspit	medieval	HA6	Chesil Street, Winchester, Hants	Wessex Arch	14
B36	*Brassica* sp./*Sinapis* sp.	seed	2143	4817	pit 4704	Iron Age	WT2	Battlesbury Bowl, Wilts	Wessex Arch	20
B34	*Aethusa cynapium* L.	fool's parsley seed	2143	4817	pit 4704	Iron Age	WT2	Battlesbury Bowl, Wilts	Wessex Arch	49
B37	*Carex* sp.-type	sedge-type seed	2119	4599	pit 4598	Iron Age	WT2	Battlesbury Bowl, Wilts	Wessex Arch	57
B38	*Carex* sp.-type	sedge-type seed	301	3045	earth floors	late medieval	KT1	Stour Street, Canterbury	Canterbury Archaeological Trust	57
B39	*Carex* sp.-type	sedge-type seed	186	2619	cesspit	late Saxon	HA1	Northgate House, Winchester, Hants	Oxford Archaeology	57
B40	*Prunus spinosa* L.	whole sloe fruit	4	28	cesspit 8	Roman	KT4	49 St Peter's Street, Canterbury, Kent	Canterbury Archaeological Trust	9, 76
B41	*Prunus domestica* L.	plum stone embedded in faecal concretion	4	28	cesspit 8	Roman	KT4	49 St Peter's Street, Canterbury, Kent	Canterbury Archaeological Trust	8

Voucher No.	Latin name	Common name & anatomical element	Sample no.	Context no.	Deposit type	Period	Site code	Site	Unit	Page in Guide
B42		folded cf. fruit skin	4	28	cesspit 8	Roman	KT4	49 St Peter's Street, Canterbury, Kent	Canterbury Archaeological Trust	76
B43	*Bupleurum rotundifolium*-type	thorow-wax seed		1049	secondary fill of cesspit	early medieval	WT4	Trowbridge, Wilts	Wessex Arch	53
B44	*Pisum sativum* L.	whole pea, little seed coat	4	28	cesspit 8	Roman	KT4	49 St Peter's Street, Canterbury, Kent	Canterbury Archaeological Trust	7
B46	*Malus* sp.	apple seed	4	28	cesspit 8	Roman	KT4	49 St Peter's Street, Canterbury, Kent	Canterbury Archaeological Trust	10
B47	*Agrimonia eupatoria* L.	agrimony fruit	221	2178	cesspit 2164	late Saxon	HA1	Discovery centre, Winchester, Hants	Oxford Archaeology	12
B48	*Anethum graveolens* L.	dill seed	245	7255	cesspit 7163	mid to late Saxon	HA5	St. Mary's Stadium, Southampton, Hants	Wessex Arch	50
B49	*Barbarea* sp./ *Sisymbrium* sp.	winter-cress/rocket seed	162	75	midden	Late Bronze Age	WT1	Potterne, Wilts	Wessex Arch	21
B51	*Lithospermum arvense* L.	field gromwell seed	2005	4124	ditch 4043	Iron Age	WT2	Battlesbury Bowl, Wilts	Wessex Arch	34
B52	*Fumaria* sp.	fumitory seed	529	1608	midden	Late Bronze Age	WT1	Potterne, Wilts	Wessex Arch	3
B54		rodent dropping cf. mouse	136	2403	cesspit 2268	mid to late Saxon	HA5	St. Mary's Stadium, Southampton, Hants	Wessex Arch	79
B55	*Rhinanthus* sp.	yellow-rattle seed			garderobe 4051	early 13th/14th C	BK1	Jennings Yard, Windsor, Berks	Wessex Arch	41
B56	*Pteridium aquilinum* (L.) Kuhn	bracken pinnule	143	372	midden	Late Bronze Age	WT1	Potterne, Wilts	Wessex Arch	1
B57	*Lamium/Ballota/ Marrubium*-type	dead-nettle/black horehound/ white horehound	529	1608	midden	Late Bronze Age	WT1	Potterne, Wilts	Wessex Arch	38
B59	*Valerianella* sp.	cornsalad seed	2194	5882	pit 4614	Iron Age	WT2	Battlesbury Bowl, Wilts	Wessex Arch	46
B60	*Myosotis* sp.	forget-me-not seed	2069	4507	pit 4458	Iron Age	WT2	Battlesbury Bowl, Wilts	Wessex Arch	35
B61	*Anagallis arvensis*-type	scarlet pimpernel-type	9	100	midden	Late Bronze Age	WT1	Potterne, Wilts	Wessex Arch	31
B62	*Anthemis/Glebionis/ Tripleurospermum* sp.-type	chamomile/corn marigold/ mayweed-type		957	primary fill of cesspit 155	early medieval	WT4	Trowbridge, Wilts	Wessex Arch	45
B63	*Ficus carica* L.	fig seed	802	841	cesspit	medieval	HA6	Chesil Street, Winchester, Hants	Wessex Arch	14
B64	*Bupleurum rotundifolium*-type	thorow-wax seed	9	1228	cesspit	Saxo-Norman	BD1	Willington to Steppingley Pipeline	Network Archaeology	53
B65	*Agrostemma githago* L.	corn cockle seed coat impression	9	1228	cesspit	Saxo-Norman	BD1	Willington to Steppingley Pipeline	Network Archaeology	27
B66	*Daucus carota* L.	carrot fruit/seed	676	3716	midden	Late Bronze Age	WT1	Potterne, Wilts	Wessex Arch	55
B67	*Hyoscyamus niger* L.	henbane seed	567	2606	midden	Late Bronze Age	WT1	Potterne, Wilts	Wessex Arch	36
B68	*Solanum nigrum* L.	black nightshade seed	333	2669	cesspit	AD1250-1400	CB1	Ferrer's Road, Huntingdon, Cambs	Oxford Archaeology	37
B69	*Solanum nigrum* L.	black nightshade seed	333	2669	cesspit	AD1250-1400	CB1	Ferrer's Road, Huntingdon, Cambs	Oxford Archaeology	37
B70	*Solanum nigrum* L.	black nightshade seed	333	2669	cesspit	AD1250-1400	CB1	Ferrer's Road, Huntingdon, Cambs	Oxford Archaeology	37
B72	*Rubus* sp.	blackberry/raspberry	333	2669	cesspit	AD1250-1400	CB1	Ferrer's Road, Huntingdon, Cambs	Oxford Archaeology	11
B73	*Rubus* sp.	blackberry/raspberry	333	2669	cesspit	AD1250-1400	CB1	Ferrer's Road, Huntingdon, Cambs	Oxford Archaeology	11
B74	*Rubus* sp.	blackberry/raspberry	333	2669	cesspit	AD1250-1400	CB1	Ferrer's Road, Huntingdon, Cambs	Oxford Archaeology	11
B76	*Aethusa cynapium* L.	fool's parsley seed	333	2669	cesspit	AD1250-1400	CB1	Ferrer's Road, Huntingdon, Cambs	Oxford Archaeology	49
B77	*Scandix pecten-veneris* L.	shepherd's needle seed	801	841	cesspit	13th C	HA6	Chesil Street, Winchester, Hants	Wessex Arch	47
B78	*Scandix pecten-veneris* L.	shepherd's needle seed	801	841	cesspit	13th C	HA6	Chesil Street, Winchester, Hants	Wessex Arch	47
B79		nodule	2162	5162	pit 5149	Iron Age	WT2	Battlesbury Bowl, Wilts	Wessex Arch	80
B80	*Chenopodium* sp./*Atriplex* sp.-type	fat hen/orache seed	2162	5162	pit 5149	Iron Age	WT2	Battlesbury Bowl, Wilts	Wessex Arch	29
B81	*Chenopodium* sp./*Atriplex* sp.-type	fat hen/orache seed	2162	5162	pit 5149	Iron Age	WT2	Battlesbury Bowl, Wilts	Wessex Arch	29
B82	*Chenopodium* sp./*Atriplex* sp.-type	fat hen/orache seed	2162	5162	pit 5149	Iron Age	WT2	Battlesbury Bowl, Wilts	Wessex Arch	29
B85	*Chenopodium* sp./*Atriplex* sp.-type	fat hen/orache seed	2162	5162	pit 5149	Iron Age	WT2	Battlesbury Bowl, Wilts	Wessex Arch	29

Voucher No.	Latin name	Common name & anatomical element	Sample no.	Context no.	Deposit type	Period	Site code	Site	Unit	Page in Guide
B86	*Ranunculus acris/bulbosus/ repens*	buttercup seed	2120	4600	pit 4598	Iron Age	WT2	Battlesbury Bowl, Wilts	Wessex Arch	4
B87	*Ranunculus acris/bulbosus/ repens*	buttercup seed	2120	4600	pit 4598	Iron Age	WT2	Battlesbury Bowl, Wilts	Wessex Arch	4
B88	*Rumex crispus*-type	curled dock-type seed	2120	4600	pit 4598	Iron Age	WT2	Battlesbury Bowl, Wilts	Wessex Arch	25
B89	*Rumex crispus*-type	curled dock-type seed	2120	4600	pit 4598	Iron Age	WT2	Battlesbury Bowl, Wilts	Wessex Arch	25
B90	*Rumex crispus*-type	curled dock-type seed	3	4	midden	Late Bronze Age-Early Iron Age	WT3	East Chisenbury, Wilts	RCHME/English Heritage (now Historic England)	25
B91	*Vicia faba* L.	field bean cotyledons	221	2178	cesspit 2164	late Saxon	HA1	Discovery Centre, Winchester, Hants	Oxford Archaeology	6
B92	*Vitis vinifera* L.	grape seed			garderobe 4051	early 13th/14th C	BK1	Jennings Yard, Windsor, Berks	Wessex Arch	5
B93	*Linum usitatissimum* L.	flax seed	333	2669	cesspit	AD1250-1400	CB1	Ferrer's Road, Huntingdon, Cambs	Oxford Archaeology	19
B94	*Linum usitatissimum* L.	flax seed	333	2669	cesspit	AD1250-1400	CB1	Ferrer's Road, Huntingdon, Cambs	Oxford Archaeology	19
B95	*Linum usitatissimum* L.	flax seed	333	2669	cesspit	AD1250-1400	CB1	Ferrer's Road, Huntingdon, Cambs	Oxford Archaeology	19
B96	*Vicia faba* L.	field bean fragment of hilum and seed coat	9	1228	cesspit	Saxo-Norman	BD1	Willington to Steppingley pipeline	Network Archaeology	6
B97	*Rhinanthus* sp.	yellow-rattle seed			garderobe 4051	early 13th/14th C	BK1	Jennings Yard, Windsor, Berks	Wessex Arch	41
B98	*Rhinanthus* sp.	yellow-rattle seed			garderobe 4051	early 13th/14th C	BK1	Jennings Yard, Windsor, Berks	Wessex Arch	41
B99	*Papaver* sp.	poppy seed	2013	4192	ditch 4043	Iron Age	WT2	Battlesbury Bowl, Wilts	Wessex Arch	2
B100	*Papaver* sp.	poppy seed	2013	4192	ditch 4043	Iron Age	WT2	Battlesbury Bowl, Wilts	Wessex Arch	2
B101	*Valerianella* sp.	cornsalad seed	2194	5882	pit 4614	Iron Age	WT2	Battlesbury Bowl, Wilts	Wessex Arch	46
B102	*Urtica dioica* L.	common nettle seed	2090	4450	ditch 4043	Iron Age	WT2	Battlesbury Bowl, Wilts	Wessex Arch	15
B103	*Urtica dioica* L.	common nettle seed	3	4	midden	Late Bronze Age-Early Iron Age	WT3	East Chisenbury, Wilts	RCHME/English Heritage (now Historic England)	15
B104	*Urtica urens* L.	small nettle seed	2103	4010	ditch 4043	Iron Age	WT2	Battlesbury Bowl, Wilts	Wessex Arch	16
B105	*Stellaria* sp.	stitchwort seed	3	4	midden	Late Bronze Age-Early Iron Age	WT3	East Chisenbury, Wilts	RCHME/English Heritage (now Historic England)	26
B108		worm cocoon	2144	4603	pit 4598	IA	WT2	Battlesbury Bowl, Wilts	Wessex Arch	78
B110	*Aphanes* sp.	parley piert seed	2144	4603	pit 4598	IA	WT2	Battlesbury Bowl, Wilts	Wessex Arch	13
B111	*Galium aparine* L.	cleavers seed	2144	4603	pit 4598	IA	WT2	Battlesbury Bowl, Wilts	Wessex Arch	33
B112	*Galium aparine* L.	cleavers seed	234	2259	pit 2258	late Saxon	HA1	Discovery centre, Winchester, Hants	Oxford Archaeology	33
B113	*Prunella vulgaris* L.	self-heal seed	555	2209	midden	Late Bronze Age	WT1	Potterne, Wilts	Wessex Arch	40
B114	*Prunella vulgaris* L.	self-heal seed	555	2209	midden	Late Bronze Age	WT1	Potterne, Wilts	Wessex Arch	40
B115	*Prunella vulgaris* L.	self-heal seed	555	2209	midden	Late Bronze Age	WT1	Potterne, Wilts	Wessex Arch	40
B116	*Poa* -type	meadow grass-type	9	100	midden	Late Bronze Age	WT1	Potterne, Wilts	Wessex Arch	58
B118	*Montia fontana* L.	blinks seed	567	2606	midden	Late Bronze Age	WT1	Potterne, Wilts	Wessex Arch	30
B119	*Montia fontana* L.	blinks seed	567	2606	midden	Late Bronze Age	WT1	Potterne, Wilts	Wessex Arch	30
B120	*Brassica* sp./ *Sinapis* sp.	cabbage/mustard seed	2193	5732	pit 5592	Iron Age	WT2	Battlesbury Bowl, Wilts	Wessex Arch	20
B121	*Brassica* sp./ *Sinapis* sp.	cabbage/mustard seed	2193	5732	pit 5592	Iron Age	WT2	Battlesbury Bowl, Wilts	Wessex Arch	20
B122	*Thlaspi arvense* L.	field penny-cress seed	2189	5730	pit 5592	Iron Age	WT2	Battlesbury Bowl, Wilts	Wessex Arch	22
B123	*Fumaria* sp.	fumitory seed	2189	5730	pit 5592	Iron Age	WT2	Battlesbury Bowl, Wilts	Wessex Arch	3
B124	*Fallopia convolvulus* (L.) Á. Löve	black bindweed fruit	301	3045	earth floors	late medieval	KT1	Stour Street, Canterbury	Canterbury Archaeological Trust	23
B125	*Hordeum vulgare* sens. lat.	barley grain	301	3045	earth floors	late medieval	KT1	Stour Street, Canterbury	Canterbury Archaeological Trust	59
B126	Poaceae	cereal/large grass culm fragment	2	108	cesspit 108	Late Iron Age/Early Roman	KT5	Church Street, Maidstone, Kent	Canterbury Archaeological Trust	73
B128	Poaceae	grass spikelet	2	108	cesspit 108	Late Iron Age/Early Roman	KT5	Church Street, Maidstone, Kent	Canterbury Archaeological Trust	73

Voucher No.	Latin name	Common name & anatomical element	Sample no.	Context no.	Deposit type	Period	Site code	Site	Unit	Page in Guide
B129	*Hordeum vulgare* sens. lat.	barley grain	801	841	cesspit	13th C	HA6	Chesil Street, Winchester, Hants	Wessex Arch	59
B130	Poaceae	culm node	801	841	cesspit	13th C	HA6	Chesil Street, Winchester, Hants	Wessex Arch	73
B131	Poaceae	culm node	801	841	cesspit	13th C	HA6	Chesil Street, Winchester, Hants	Wessex Arch	73
B132	*Ficus carica* L.	fig fruit	801	841	cesspit	13th C	HA6	Chesil Street, Winchester, Hants	Wessex Arch	14
B133	*Galeopsis* sp.	hemp-nettle fruit	555	2209	midden	Late Bronze Age	WT1	Potterne, Wilts	Wessex Arch	39
B134	*Galeopsis* sp.	hemp-nettle fruit	3	4	midden	Late Bronze Age-Early Iron Age	WT3	East Chisenbury, Wilts	Royal Commission? McOmish, Field & Brown	39
B136	*Coriandrum sativum* L.	coriander fruit pericarp	10008	10114	cesspit 10116	mid Saxon	HA5	St. Mary's Stadium, Southampton, Hants	Wessex Arch	48
B137	*Coriandrum sativum* L.	coriander seed	10008	10114	cesspit 10116	mid Saxon	HA5	St. Mary's Stadium, Southampton, Hants	Wessex Arch	48
B138	*Viola* sp.	violet or pansy seed	2162	5162	pit 5149	Iron Age	WT2	Battlesbury Bowl, Wilts	Wessex Arch	17
B139	*Eleocharis* sp.	spike-rush fruit	573	2777	midden	Late Bronze Age	WT1	Potterne, Wilts	Wessex Arch	56
B140	*Eleocharis* sp.	spike-rush seed	537	1810	midden	Late Bronze Age	WT1	Potterne, Wilts	Wessex Arch	56
B142	*Torilis* sp.	hedge-parsley fruit	354	6137	pit 6138	late Saxon	HA1	Northgate House, Winchester, Hants	Oxford Archaeology	54
B143	*Pteridium aquilinum* (L.) Kuhn	bracken pinnule, lower surface	2	108	cesspit 108	Late Iron Age/Early Roman	KT5	Church Street, Maidstone, Kent	Canterbury Archaeological Trust	1
B144	*Conium maculatum* L.	hemlock seed			base of City Ditch	13th-16th C	LD2	London City Ditch, Aldersgate	Pre-Construct Archaeology	52
B146	*Sherardia arvensis* L.	field madder seed	2120	4600	pit 4598	Iron Age	WT2	Battlesbury Bowl, Wilts	Wessex Arch	32
B147	Poaceae	bran fragments in concretion	801?	841	cesspit	13th C	HA6	Chesil Street, Winchester, Hants	Wessex Arch	72
B148	*Viola* sp.	violet or pansy seed	801	841	cesspit	13th C	HA6	Chesil Street, Winchester, Hants	Wessex Arch	17
B149	Poaceae	matted straw/grass culms	801	841	cesspit	13th C	HA6	Chesil Street, Winchester, Hants	Wessex Arch	61
B150	*Centaurea cyanus* L.	cornflower seed	802	841	cesspit	medieval	HA6	Chesil Street, Winchester, Hants	Wessex Arch	43
B151	*Centaurea cyanus* L.	cornflower fruit			garderobe 4051	early 13th/14th C	BK1	Jennings Yard, Windsor, Berks	Wessex Arch	43
B152	*Carduus* sp./*Cirsium* sp.	thistle-type seed	2120	4600	pit 4598	Iron Age	WT2	Battlesbury Bowl, Wilts	Wessex Arch	42
B153	*Carduus* sp./*Cirsium* sp.	thistle-type seed	2120	4600	pit 4598	Iron Age	WT2	Battlesbury Bowl, Wilts	Wessex Arch	42
B154	*Silene* sp.	campion seed	9	21	midden	Late Bronze Age-Early Iron Age	WT3	East Chisenbury, Wilts	Royal Commission? McOmish, Field & Brown	28
B155	*Anethum graveolens* L.	dill seed	245	7255	cesspit 7163	mid to late Saxon	HA5	St. Mary's Stadium, Southampton, Hants	Wessex Arch	50
B156	*Barbarea* sp./*Sisymbrium* sp.	winter-cress/rocket seed	9	100	midden	Late Bronze Age	WT1	Potterne, Wilts	Wessex Arch	21
B157	*Pteridium aquilinum* (L.) Kuhn	bracken pinnule fragment	234	2259	pit 2258	late Saxon	HA1	Discovery centre, Winchester, Hants	Oxford Archaeology	1
B158		rodent dropping	136	2403	cesspit 2268	mid to late Saxon	HA5	St. Mary's Stadium, Southampton, Hants	Wessex Arch	79
B159	*Foeniculum vulgare* L.	fennel seed	1	167	cesspit D	post-medieval	LD1	Cockspur Street, London	Pre-Construct Archaeology	51
B160	*Foeniculum vulgare* L.	fennel seed	1	167	cesspit D	post-medieval	LD1	Cockspur Street, London	Pre-Construct Archaeology	51
B161	*Pisum* sp./*Vicia* sp.	pulse seed coat				unstratified	HA1	Discovery centre, Winchester, Hants	Oxford Archaeology	75
B162	*Pisum* sp./*Vicia* sp.	pulse seed coat	180	2377	pit 2373	late Saxon	HA1	Northgate House, Winchester, Hants	Oxford Archaeology	75
B163	*Malus* sp./*Pyrus* sp.	apple/pear seed	101	1087	pit 1088	early medieval	KT1	Stour Street, Canterbury	Canterbury Archaeological Trust	10
B164	*Prunus domestica* L.	plum-type kernel	257	2342		Anglo-Norman	HA1	Discovery centre, Winchester, Hants	Oxford Archaeology	8
B165	Poaceae	bran curl	9	1228	cesspit	Saxo-Norman	BD1	Willington to Steppingley pipeline	Network Archaeology	72
B166	*Aethusa cynapium* L.	fool's parsley seed	2143	4817	pit 4704	Iron Age	WT2	Battlesbury Bowl, Wilts	Wessex Arch	49
B167	*Malus* sp./*Pyrus* sp.	apple/pear seed	3	4	midden	Late Bronze Age-Early Iron Age	WT3	East Chisenbury, Wilts	RCHME/English Heritage (now Historic England)	10

Voucher No.	Latin name	Common name & anatomical element	Sample no.	Context no.	Deposit type	Period	Site code	Site	Unit	Page in Guide
B168	*Prunus spinosa* L.	sloe kernel	4	28	cesspit 8	Roman	KT4	49 St Peter's Street, Canterbury, Kent	Canterbury Archaeological Trust	9
B169	*Myosotis* sp.	forget-me-not seed	2201	5715	pit 5670	Iron Age	WT2	Battlesbury Bowl, Wilts	Wessex Arch	35
B171	*Lamium/Ballota/ Marrubium*-type	dead-nettle/black horehound/ white horehound	529	1608	midden	Late Bronze Age	WT1	Potterne, Wilts	Wessex Arch	38
B172	*Anthemis/Glebionis/ Tripleurospermum*-type	chamomile/corn marigold/ mayweed-type		957	primary fill of cesspit 155	early medieval	WT4	Trowbridge, Wilts	Wessex Arch	45
B173	*Anthemis/Glebionis/ Tripleurospermum*-type	chamomile/corn marigold/ mayweed-type		957	primary fill of cesspit 155	early medieval	WT4	Trowbridge, Wilts	Wessex Arch	45
B174	*Anthemis/Glebionis/ Tripleurospermum*-type	chamomile/corn marigold/ mayweed-type	2136	4728	pit 4641	Iron Age	WT2	Battlesbury Bowl, Wilts	Wessex Arch	45
B175	*Juncus* sp.	rush stem	180	2377	pit 2373	late Saxon	HA1	Northgate House, Winchester, Hants	Oxford Archaeology	74
B176		dicotyledonous root	4	50	midden	Late Bronze Age	WT1	Potterne, Wilts	Wessex Arch	77
E1	*Psychoda alternata* (Say)	puparium	7122	3345	pit (probably cesspit)	12th to 13th C AD	BR2	Finzel's Reach Bristol	Oxford Archaeology	60
E2	*Eristalis tenax* (L.)	puparium	320	3488	pit fill	Roman	LR7	Vine Street, Leicester	University of Leicester Archaeology Service	61
E3	*Sepsis spp.*	puparium	3331	7068	pit fill	12th to 13th C AD	BR2	Finzel's Reach Bristol	Oxford Archaeology	62
E4	*Thoracochaeta zosterae* (Hal.) (light forms)	puparium	259	5992	privy fill	medieval	LR6?	Freeschool Lane, Leicester	University of Leicester Archaeology Service	63
E4	*Thoracochaeta zosterae* (Hal.) (Dark forms)	puparium	7122	3345	pit (probably cesspit)	12th to 13th C AD	BR2	Finzel's Reach Bristol	Oxford Archaeology	63
E5	*Calliphora vomitaria* (L.)	puparium	-	-	modern specimen			University of Birmingham		64
E6	*Fannia ?scalaris* (L.)	puparium	320	3488	pit fill	roman	LR7	Vine Street, Leicester.	University of Leicester Archaeology Service	65
E7	*Musca domestica* (L.)	puparium	5	115	peaty layer	medieval	HE1	14-19 Bridge Street, Hereford	Hereford County Archaeology Service	66
E8	*Muscina stabulans* (Fall.)	puparium	6	94	mortar lined pit	medieval	HE1	14-19 Bridge Street, Hereford	Hereford County Archaeology Service	67
E9	*Stomoyxs calcitrans* (L.)	puparium	3380	7487	fill of barrel lined pit	14th to 15th C AD	BR2	Finzel's Reach Bristol	Oxford Archaeology	68
E10	*Melophagus ovinus* (L.)	puparium	3176	4432	pit	13th-14th C AD	BR2	Finzel's Reach Bristol	Oxford Archaeology	69
E11	*Oryzaephilus surinamnesis* (L.)	head, thorax and elytra of adult	4	86	timber lined well	Roman 2nd C AD	WL1	Inveresk Gate, West Lothian	AOC Archaeology	70
E12	*Sitophilus granarius* (L.)	head, thorax and elytra of adult	4	86	timber lined well	Roman 2nd C AD	WL1	Inveresk Gate, West Lothian	AOC Archaeology	71

Appendix II: Archaeological sites

Site Code	Site Location	Reference
BD1	Willington to Steppingley Pipeline, Beds	Carruthers, W J unpublished 2003 'Charred, Mineralised and Waterlogged Plant Remains Report', *in* Anon, Willington to Steppingley 900mm Gas Pipeline Archaeological Evaluation, Excavation and Watching Brief 2002. Network Archaeology Ltd Report Volume 3; Appendix 12. Archaeobotanical Report. ADS https://doi.org/10.5284/1021699
BK1	Jennings Yard, Windsor, Berks	Carruthers, W J 1993 'Carbonised, mineralised and waterlogged plant remains', *in* Hawkes, J W and Heaton, M J, *Jennings Yard, Windsor: A Closed-Shaft Garderobe and Associated Medieval Structures. Wessex Archaeology Report* **3**, 82-90
BR1	45-53 West Street, Bedminster, Bristol	Griffiths, C J 2014-15 'Plant remains', *in* Young, D and Young, A, Archaeological Excavations at 45-53 West Street, Bedminster, Bristol, 2005. *Bristol & Avon Archaeology* **26**, 88-90
BR2	Finzel's Reach, Bristol	Smith, D N 2017 'The insect remains', in Ford, B M, Brady, K and Teague, S (eds) *From Bridgehead to Brewery: the medieval and post-medieval remains from Finzel's Reach, Bristol.* Oxford Archaeology Monograph **27**. Oxford archaeology 274-276. Section 22.23 on DVD
CB1	Ferrar's Road, Huntingdon, Cambs	Carruthers, W J (in preparation) 'Mineralised plant remains', *in* Land between Ferrar's Road, Dryden's Walk and Edison Bell Way, Huntingdon, Cambs., for Oxford Archaeology.
CB2	Grand Arcade & Bradwell's Court, Cambridge, Cambs	Smith, D 'The insects' and Ballantyne, R M and de Vareilles, A K 'Plant macrofossils', *in* Cessford, C and Dickens, *A forthcoming Medieval to modern suburban material culture and sequence at Grand Arcade, Cambridge: archaeological investigations of an eleventh to twentieth-century suburb and town ditch.* McDonald Institute Monograph/Cambridge Archaeology Unit Urban Archaeology Series, The Archaeology of Cambridge Volume 1
HA1	Northgate House and Discovery Centre, Winchester, Hants	Carruthers, W 2011 'Charred and mineralised plant remains', *in* Ford, B M and Teague, S *Winchester - a City in the Making: Archaeological excavations between 2002 and 2007 on the sites of Northgate House, Staple Gardens and the former Winchester Library, Jewry St.* Oxford Archaeology Monograph **12**, 363-373
HA2	Silchester, Hants	Robinson, M, Fulford, N and Tootell, K 2006 'Chapter 5: The Macroscopic Plant Remains', *in* Fulford, M, Clarke, A and Eckardt, H *Life and Labour in Late Roman Silchester. Excavations in Insula IX since 1997.* Britannia Monograph **22**, 206-380
HA3	Flint Farm, Hants	Campbell, G 2008 'Charred and mineralised plant remains', *in* Cunliffe, B and Poole, C *The Danebury Environs Programme: A Wessex landscape in the Roman era Volume 2-part 6: Flint Farm, Goodworth Clatford, Hants, 2004* English Heritage and Oxford University School of Archaeology Monograph **71**, 99-100 and e text http://projects.arch.ox.ac.uk/DERP.html
HA4	The Deanery, Southampton, Hants	Pelling, R 2012 'Mineralised and charred plant remains', *in* Birbeck, V, Middle Saxon Settlement at the Deanery, Chapel Road, Southampton. *Proceedings of the Hampshire Field Club and Archaeology Society* **67:2** 308-311 (Hampshire Studies 2012)
HA5	St Mary's Stadium, Southampton, Hants	Carruthers, W J 2005 'Mineralised Plant Remains', *in* Birbeck, V with Smith, RJC, Andrews, P and Stoodley, N *The Origins of Mid-Saxon Southampton: Excavations at the Friends Provident St Mary's Stadium 1998-2000*, Wessex Archaeology Report, 157-163, 183-184
HA6	Chesil Street, Winchester	Carruthers, W and López-Dóriga, I forthcoming 'Charred and mineralised plant remains from medieval cess pit 842', *in* Orczewski, P and Andrews, P, Romano-British and medieval extra-mural settlement at Chesil Street, Winchester. Hampshire Studies.
HA7	Abbots Worthy, Hampshire	Carruthers, W J 1991 'The plant remains', *in* Fasham P J and Whinney, R J B *Archaeology and the M3.* Hampshire Field Club Monograph **7**, (with Trust for Wessex Archaeology) 67-75
HA8	French Quarter, Southampton	Smith, W 2011 'Plant remains', *in* Brown, R and Hardy, A (eds) *Trade and Prosperity, War and Poverty: An Archaeological and Historical Investigation into Southampton's French Quarter.* Oxford Archaeology Monograph **15**, 243-251. Oxford: Oxford Archaeology. Available online at https://library.thehumanjourney.net/47/1/SOU_1382_Specialist_report_download_E4.pdf (accessed 15/10/19)
HA9	Anderson's Road, Southampton	Stevens, C 2005 'Charred, mineralised and waterlogged plant remains', *in* Ellis, C and Andrews, P *A Mid-Saxon Site at Anderson's Road, Southampton.* Wessex Archaeology Report (for publication in Hampshire Studies) 24-28. Available online at https://archaeologydataservice.ac.uk/archiveDS/archiveDownload?t=arch-1184-1/dissemination/pdf/Publication_text_and_illustrations/SOU1240_Andersons_road_publication_report.pdf (accessed 15/10/19)
HE1	Bridge Street, Hereford	Smith, D N and Skidmore, P unpublished 1999 'The insect remains from 14-19 Bridge Street, Hereford.' Report to Hereford County Archaeology Unit
KT1	Stour Street, Canterbury	Carruthers, W and Allison, E 2015 'Plant and insect remains from medieval features at 70 Stour Street, Canterbury, Kent (Site Code SSC(70). EX13).' *Canterbury Archaeological Trust Report* **2015/79**, July 2015,
KT2	St Lawrence Cricket Ground, Canterbury	Smith, W unpublished script 2015 'Charred and Mineralised Plant Macrofossils from prehistoric, medieval and post-medieval features at the Bat and Ball Site, St. Lawrence Cricket Ground, Canterbury.' Canterbury Archaeological Trust

Site Code	Site Location	Reference
KT4	49, St Peter's Street, Canterbury, Kent	Carruthers, W J and Allison E 2015 49 St Peter's Street, Canterbury (49SPS.EX15): The plant and insect remains, Canterbury Archaeological Trust Report 2015/177, November 2015
KT5	Church Street Maidstone	Carruthers, W J 2014 'The Charred and Mineralised Plant Remains', *in* O'Shea, L and Weeks, J, Evidence of a distinct focus of a Romano-British settlement at Maidstone? Excavations at Church Street 2011-12. *Archaeologia Cantiana* **135**, 143-147
LD1	Cockspur Street, London	Pickard, C 2002 Excavations at 25-34 Cockspur Street and 6-8 Spring Gardens. *London Archaeologist* **10:02**, 31-40 Available online at https://archaeologydataservice.ac.uk/archiveDS/archiveDownload?t=arch-457-1/dissemination/pdf/vol10/vol10_02/10_02_031_040.pdf (accessed 15/10/19)
LD2	London City Ditch, Aldersgate	Carruthers, W 2001 'The Charred Plant Remains', *in* Butler, J The City Defences at Aldersgate. *LAMAS Transactions* **52**, 99-106.
LR1	Elms Farm, Humberstone, Leicester, Leics	Pelling, R 2000 'The charred and mineralised plant remains', *in* Charles, B M, Parkinson, A and Foreman, S, A Bronze Age ditch and Iron Age Settlement at Elms Farm, Humberstone, Leicester. *Transactions of the Leicestershire Archaeology and History Society* **74**, 207-213
LR2	Causeway Lane, Leicester, Leics	Monckton, A 1999 'The Plant Remains', *in* Connor, A and Buckley, R *Roman and Medieval Occupation in Causeway Lane, Leicester*. Leicester Archaeology Monograph **5**, pp.346-362.
LR4	Bonners Lane, Leicester, Leics	Monckton, A 2004 'Plant macrofossils', *in* Finn, N *The Origins of a Leicester Suburb: Roman, Anglo-Saxon, medieval and post-medieval occupation on Bonners Lane*. British Archaeological Reports British Series **372**, 156-166
LR5	St Nicholas Place, Leicester, Leics	Monckton, A and Boyer, P unpublished 2008 Leicester, 9 St. Nicholas Place and Medieval Undercroft: charred and mineralised plant remains from excavations in 1989 and 2003. University of Leicester Archaeology Service Report 2009-110.
LR6	Freeschool Lane, Leicester, Leics	Radini, A 2009 'The Plant Remains from Freeschool Lane, Leicester', *in* Coward, J and Speed, G, Excavations at Freeschool Lane, Leicester, Highcross Project. University of Leicester Archaeology Service Report 2009-140.
LR7	Vine Street, Leicester, Leics	Smith, D N unpublished 2009 *Fly puparia from Vine Street (A24.2003: A22.2003) and Freeschool Lane*, (A8.2005) Leicester. University of Birmingham Environmental Archaeology Services report 173
NK1	Castle Acre Castle, Norfolk	Green, F J 1982 'The Plant Remains', *in* Coad J G and Streeten A D F Excavations at Castle Acre, Norfolk 1972-77, Country House and Castle of the Norwich Earls of Surrey. *Archaeological Journal* **139**, 273-275
OX1	Ashmolean Museum, Oxford, Oxon	Smith, W forthcoming Section within Chapter 11 'Medieval and post-medieval tenements at the Ashmolean Museum Extension Site, by Teague, S and Ford BM *in* 'The Archaeology of Oxford in the Twenty-first Century: Investigations in the Historic City and Northern Suburb by Oxford Archaeology 2006–16.' Dodd, A, Mileson, S and Webley, L (eds) *Oxoniensia*
WL1	Inveresk Gate, East Lothian	Smith, D N 2004 'The insect remains from the well', *in* Bishop, M C *Inveresk Gate: Excavations in the Roman Civil Settlement at Inveresk, East Lothian, 1996–2000*. STAR Monograph **7**. Loanhead, Midlothian: Scottish Trust for Archaeological Research 81-88
WT1	Potterne, Wilts	Carruthers, W J 2000 'The mineralised plant remains', *in* Lawson, A J and Gingell C J, *Potterne 1982-5: animal husbandry in later prehistoric Wiltshire*. Wessex Arch. Rep. **17**, 72-84, 91-95
WT2	Battlesbury Bowl, Wilts	Carruthers, W 2008 'Mineralised plant remains', *in* Ellis, C and Powell, A B with Hawkes, J, *An Iron Age Settlement outside Battlesbury Hillfort, Warminster, and Sites along the Southern Range Road*. Wessex Archaeology Report **22**, 102-114
WT3	East Chisenbury, Wilts	Carruthers, W J unpublished draft 2008 'The plant remains', *in* McOmish, D, Field, D and Brown, G, The Late Bronze Age - Early Iron Age Midden Site at East Chisenbury, Wiltshire.
WT4	Trowbridge, Wilts	Carruthers, W J 1993 'Carbonised, mineralised and waterlogged plant remains from cess pit F155', *in* Graham, A H and Davis, S M, *Excavations in Trowbridge, Wiltshire 1977 and 1986-88*. Wessex Archaeology Report **2**, 136-141

Glossary - Botanical

Apical tubercle – (in *Eleocharis* sp.) expanded and persistent base of style, useful for identification purposes if preserved.

Apomixis – a type of asexual reproduction found in angiosperms not involving fertilisation, eg the replacement of seed by a plantlet or production of bulbils.

Archaeophyte –an alien plant which was introduced and became naturalised in a study area before 1500 CE.

Campylotropous – curved or folded embryo or seed.

Chalaza scutellum – (in *Vitis vinifera*) oval structure on dorsal side of grape seed, the form used for identification purposes.

Fossette – (in *Vitis vinifera*) a depression, in *Vitis* found either site of the ridge (raphe) on the ventral side, the form being useful for purposes of identification.

Globose – spherical.

Hilum – scar on seed where it was attached to ovary wall or funiculus.

Hypanthium – (in *Agrimonia* sp.) cup-shaped structure formed by extension of receptacle above base of ovary.

Mericarp – a one-seeded portion formed by the splitting up of a 2 to many-seeded fruit, i.e. the seed-dispersal unit.

Micropyle – small opening of the integument(s), located at the end of the hilum in Fabaceae.

Pericarp – part of the fruit that is formed from the wall of the ripening ovary, differentiated into distinctive layers in some fruits eg stone fruits (drupes) but not in others, eg Apiaceae.

Perisperm – diploid storage tissue in angiosperms that surrounds the embryo of certain seeds, usually storing starch.

Pinnule – (mainly used in ferns) smallest lobe of bi- (or more) pinnate leaf.

Raphe – (in *Vitis vinifera*) prominent longitudinal ridge on ventral side of seed where two halves of embryo fused.

Reniform – kidney-shaped.

Reticulate – forming a network.

Spinose – having spines, spiny.

Stylopodium – (in Apiaceae) prominent nectar-secreting disc and swollen style base present at apex of fruits of sub-family Apioideae.

Terete – rounded in section.

Tubercle – small spherical or ellipsoid swelling, **tuberculate** fruits and seeds having their outer surfaces covered with many small tubercles.

Verrucose – surface covered in small wart-like outgrowths.

Vitta, (vittae) – (in Apiaceae) oil canal(s) persisting in most fruits of sub-families Apioideae and Saniculoideae.

Glossary – Entomological

'belts' 'ambulatory welts' - roughened areas of cuticle on the surface of the puparium that were used to move or 'shuffle along' when the fly was a maggot. Often associated with the join between the sections of the maggot or puparium.

Bifurcated - the shape of processes that divide into two forks from a single stem.

Cuticle - the surface and material that make up the exoskeleton of all insects. It is normally made of chitin and in the surface of puparia and Coleoptera (beetles) can be heavily sclerotised to form a 'shell'.

Elytra - the plural of elytron. Elytra are the heavily chitinized and hardened 'shells' or 'wing cases' that cover the wings of beetles and some bugs. In other insects where these are not so strongly chitinized these are the front pair of wings.

Processes - any spike, tubular or 'feathery' structures that are part of the exoskeleton of the insect.

Puparium - the hardened exoskeleton of the last larval instar (the maggot). The anterior end is normally the thinner wedge-shaped end which bears the mouth parts. The posterior end is normally wider and flatter and often bears the two 'eye' like posterior spiracles.

Rostrum - the long snout-like projection on the front of the head of weevils. This bears the mouth parts and the antennae of the insect.

Setae - hair like sensory organs which project from the exoskeleton. These can be preserved on the surface of puparia as 'spikes' or 'hairs'.

Spiracles - external openings in the cuticle or exoskeleton of the insect through which it breaths and which lead to the internal respiratory system of the insect.

Thorax - in Coleoptera (beetles) the body of the adult is divided into three sections. The thorax is the middle of these main three sections and consists of three segments each bearing a pair of legs. The last two also bear wings.

References

Belshaw, R 1989 'A note on the recovery of *Thoracochaeta zosterae* (Haliday) (Diptera: Sphaeroceridae) from archaeological deposits.' *Circaea* **6**, 39–41

Berggren, G 1981 *Atlas of seeds and small fruits of Northwest-European plant species. Part 3. Salicaceae-Cruciferae.* Berlings, Arlöv, Sweden

Buckand, PC and Perry, DW 1989 'Ectoparasites of sheep from Storaborg, Iceland and their interpretation: piss, parasites and people, a palaeoecological perspective'. *Hikuin* **15**, 37-46

Cappers, RTJ, Bekker, RM and Jans, JEA 2006 *Digitale Zadenatlas van Nederland. Digital seed atlas of the Netherlands.* Groningen, Netherlands: Barkhuis Publishing and Groningen University Library

Carruthers, W 1988 'Mystery object no. 2 – animal, mineral or vegetable?' *Circaea* **6**, 20

Carruthers, W J 2000 'The mineralised plant remains' in Lawson, A J and Gingell, C J *Potterne 1982-5: animal husbandry in later prehistoric Wiltshire.* Wessex Arch. Rep. **17**, 72-84, 91-95

Corner, EJH 2009 *The seeds of dicotyledons.* Cambridge: Cambridge University Press. Digitally printed version, first published 1976

Ellenberg, H 1988 *Vegetation ecology of Central Europe.* Cambridge: Cambridge University Press

Fritsch, R 1979 'Zur Samenmorphologie des Kulturmohns (*Papaver somniferum* L.)'. *Die Kulturpflanze* **27(2)** 217-227

Girling, MA 1984 'Eighteenth century records of human lice (Pthiraptera, Anoplura) and fleas (Siphoaptera, Pulicidae) in the City of London.' *Entomologist's Monthly Magazine* **120**, 207-210

Green, F 1979 'Phosphate mineralisation of seeds from archaeological sites.' *Journal of Archaeological Science* **6**, 279-284

Greig, J 1991 'The early history of the cornflower (*Centaurea cyanus* L.) in the British Isles' *in* Eva Hajnalová (ed) *Palaeoethnobotany and Archaeology, International Work Group for Palaeoethnobotany 8th Symposium, Nitra-Nové Vozokany 1989* Acta Interdisciplinaria Archaeologica **VII**, 97-109

Greig-Smith, PW 1948 'Biological flora of the British Isles, No. 23, *Urtica dioica* L'. *Journal of Ecology* **36**, 343-351

Grieve, Mrs M 1992 *A modern herbal: The medicinal, culinary, cosmetic and economic properties, cultivation and folklore of herbs, grasses, fungi, shrubs and trees with all their modern scientific uses.* London: Tiger Books International. Reprint of revised edition, first published in 1931 by Jonathan Cape Ltd

Hall, A and Kenward, H 2003 'Can we identify biological indicator groups for craft, industry and other activities?' *in* Murphy, P and Wiltshire P E J (eds) *The environmental archaeology of industry.* Symposia of the Association for Environmental Archaeology **20**. Oxford: Oxbow, 114-30

Hill, MO, Mountford, JO, Roy, DB and Bunce, RG H 1999 *Ellenberg's indicator values for British plants.* ECOFACT Volume 2: Technical Annex. Institute of Terrestrial Ecology

Kenward, HK 1999 'Pubic lice (*Pthirus pubis* L.) were present in Roman and Medieval Britain.' *Antiquity* **73**, 911-915

Kenward, HK and Hall, AR 1997 'Enhancing bio-archaeological interpretation using indicator groups: Stable manure as a paradigm'. *Journal of Archaeological Science* **24,** 663–673

McCobb, LME, Briggs, DEG, Evershed, RP and Hall, RA 2001 'Preservation of fossil seeds From a 10th century AD cess pit at Coppergate, York.' *Journal of Archaeological Science* **28**, 929-940

McCobb, LME, Briggs, DEG, Carruthers, WJ and Evershed, RP 2003 'Phosphatisation of seeds and roots in a Late Bronze Age deposit at Potterne, Wiltshire, UK.' *Journal of Archaeological Science* **30**, 1269-1281

Panagiotakopulu, E and Buckland, PC 2017 'A thousand bites - Insect introductions and late Holocene environments'. *Quaternary Science Reviews* **156**, 23- 35

Phillips, R 1983 *Wild food.* Pan Books

Phipps, J 1988 'Mystery object identified.' *Circaea* **5**, 58-60

Piearce, TG, Oates, K and Carruthers, WJ 1992 'Fossil earthworm cocoons from a Bronze Age site in Wiltshire, England.' *Soil Biology and Biochemistry* **24 Issue 12**, 1255-1258

Pigott, CD and Taylor, K 1964 'The distribution of some woodland herbs in relation to the supply of nitrogen and phosphorous in the soil'. *Journal of Ecology* **52**, 175-185

Radini, A 2009 'The plant remains from Freeschool Lane, Leicester' *in* Coward, J and Speed, G *Excavations at Freeschool Lane, Leicester, Highcross Project.*

(Unpublished report for University of Leicester Archaeological Services, Report 2009-140)

Robinson, M, Fulford, N, and Tootell, K 2006 'Chapter 5: The macroscopic plant remains' *in* Fulford, M Clarke, A and Eckardt, H *Life and labour in Late Roman Silchester. Excavations in Insula IX since 1997.* Britannia Monograph **22**, 206-380

Skidmore, P 1999 'The Diptera' *in* Connor, A and Buckley, R (eds) *Roman and Medieval Occupation in Causeway Lane, Leicester* (Leicester Archaeological Monograph **5**), 341–343. Leicester: Leicester University Press

Smith, DN 2008 *Fly pupae from Vine Street (A24.2003; A22.2003), and Freeschool Lane, (A8.2005) Leicester.* (Unpublished report to University of Leicester Archaeological Service)

Smith, DN 2009 *Mineralised and waterlogged fly pupae, and other insects and arthropods Southampton French Quarter 1382* (Specialist Report Download E9). Oxford: Oxford Archaeology http://library.thehumanjourney. net/52/1/SOU_1382_Specialist_report_download_ E9.pdf (accessed 15/10/19)

Smith, DN 2011 'Insects from Northfleet' *in* Barnett, C, McKinley, J I, Stafford, E, Grimm, J M and Stevens, C J (eds) *Settling the Ebbsfleet Valley: High Speed 1 Excavations at Springhead and Northfleet, Kent, The late Iron Age, Roman, Saxon and Medieval landscape* (*Volume 3: Late Iron Age to Roman Human remains and Environmental Reports*), 88–90. Oxford/ Salisbury: Oxford Wessex Archaeology

Smith, DN 2012 *Insects in the city: An Archaeoentomological Perspective on London's Past.* British Archaeological Reports British Series **561**. Oxford: Archaeopress

Smith, DN 2013 Defining an 'indicator package' to allow identification of 'cess pits' in the archaeological record. *Journal of Archaeological Science* **40**, 526–43

Smith, DN 2017 'The insect remains' *in* Ford, B M, Brady, K and Teague, S (eds) *From Bridgehead to Brewery: the medieval and post-medieval remains from Finzel's Reach, Bristol.* Oxford Archaeology Monography **27**, 274-276. Oxford Archaeology

Smith, DN and Kenward HK 2013 'Well, Sextus, what can we do with this?' The disposal and use of insect-infested grain in Roman Britain. *Environmental Archaeology* **17**, 141–50

Smith, DN and Kenward, HK 2011 Roman grain pests in Britain: Implications for grain supply and agricultural production. *Britannia* **42**, 243–62

Smith, KGV 1973 *Insects and other arthropods of medical importance.* London: British Museum (Natural History)

Smith, KGV 1989 *An introduction to the immature stages of British flies.* (Handbooks for the identification of British insects X, 14). London: Royal Entomological Society of London

Stace, C 2010 *New Flora of the British Isles.* Third Edition. Cambridge: Cambridge University Press

Stuart, M 1987 *The encyclopedia of herbs and herbalism.* London: Black Cat

Stuppy, W 2004 *Glossary of seed and fruit morphological terms.* Seed Conservation Department, Royal Botanic Gardens

Treasure, ER and Church, MJ 2017 'Can't find a pulse? Celtic bean (*Vicia faba* L.) in British prehistory.' *Environmental Archaeology.* **22**, 113-127

Tutin, TG 1980 *Umbellifers of the British Isles.* Botanical Society of the British Isles Handbook **No. 2**

van der Veen, M, Livarda, A and Hill, A 2008, 'New plant foods in Roman Britain – dispersal and social access.' *Environmental Archaeology* **13:1** 11-36

Watson, L and Dallwitz, MJ *The families of flowering plants.* (Available online at http://www.delta-intkey. com/angio/www/urticace.htm) (accessed 16/10.19)

Webb SC, Hedges REM and Robinson M 1998 'The seaweed fly *Thoracochaeta zosterae* (Hal.) (Diptera: Sphaerocidae) in inland archaeological contexts: ^{13}C and ^{15}N solves the puzzle.' Journal of Archaeological Science **25**, 1253-1257

Zohary, D, Hopf, M and Weiss, E 2013 *Domestication of Plants in the Old World.* Oxford: Oxford University Press, 4th edition

Online Sources

Web 1: Plants for a Future (https://pfaf.org/user/ Default.aspx) (accessed 16/10/19)

Web 2: The Online Atlas of the British and Irish Flora, Biological Records Centre (www. brc.ac.uk/plantatlas/) (accessed 16/10/19)

Web 3: Drugs.com: Prescription Drug Information, Interactions and Side Effects (https://www.drugs.com/ npc/) (accessed 16/10/19)

Index

Printed and bound by CPI Group (UK) Ltd, Croydon, CR0 4YY

16/04/2025

14658580-0001